System of Rational Discourse — Volume 4

PROPOSITIONS

JOHN TROTTER

Just Talk

ISBN 0 9587584 3 3

Just Talk Pty Ltd, 70 Barada Crescent, Aranda, ACT, Australia

PREFACE

In this volume, I develop the theory of intensions further and connect SRD with more and more of ordinary English. I show (surprise, surprise!) that ordinary language provides a perfect instrument for intensional (indeed any) discourse, if it is used carefully and systematically. Many authors (not professional logicans) have used care, but they have not worked according to any system that they could specify publicly in detail or that is precise enough to prevent their own mistakes.

The reason that such a development has not been published before now is complicated. First, the logical calculi conventionally employed have not been sufficiently versatile, so that intuitively obvious results have not come through, as they do here. Then the weight of authority and prestige has rested with those who value precision above sense, and it still does. As a result, pseudo-problems stemming from well specified but ill interpreted formulas have been taken at face value rather than being treated as signs of logical trouble. So a modern intellectual is faced with the choice, which is decided intuitively without realization of it, of working rigorously with amorphous subject matter or talking imprecisely about *perhaps* well specified subject matter. Further, people have let themselves "go gentle into that goodnight", not being intellectually robust enough to "rage, rage against the dying of the light". Quine knows what I mean, but he has drawn the line in a different place. I have devoted a solitary career to a self-imposed task that no one else seems to see. Incomprehension by a well meaning establishment has promoted parochial obstruction leading to global neglect. Finally, technological development has not till now provided inexpensive means of publishing material that conservative editors, mindful of their readers' interests, have shown no interest in. It would be naive and historically ignorant to expect any different response from an establishment. My "philosophical" reactions have been set out polemically in *The Emperors' New Logic*.

As in previous volumes, discussion here in unformalized ordinary language carries no philosophical weight. As the exposition proceeds, the calculus provides a precise account of doctrine suggested or outlined in informal discussion. Attributes (intensions) but not predicates (notations) have already been formally treated. Here predications (intensions) but not sentences (notations) are dealt with. The strategy of the whole SRD project is to treat meanings primarily and their expressive notations derivatively, in direct opposition to the dominant contemporary strategy.

As a pedagogical device, explanatory discussion here will involve notations, so that problems may be presented in a familiar context. But formal discussion will treat only intensions. Discussion of notations will be treated rigorously in Volume 5.

Sub-editing is seldom perfect, and occasional lapses do not warrant global condemnation. If the standard here falls below what is normally acceptable, I offer my apologies. An author is never a good choice for editing his own work, but in my position I have no alternative.

CONTENTS

Those who strain at the gnat of meanings
typically swallow a camel of ineffective substitutes.

INTRODUCTION

The present position

Items (1) to (5) of <u>1</u>.A4.1 (in Volume <u>1</u>) provide a recursive specification of primitive attributes, presented by a user's subject matter, and Boolean compounds of them. But the specification of attributes remains incomplete. What I have taken as attributes for entailment and for definite descriptions have been extensively examined without rigorous proof that they are indeed attributes. The reason is that their definientia involve conditions, such as the quantified condition in 1.5.6D,

1.5.6D $B \Rrightarrow y$ = [it is *predle* & $(u)(\sim u\ E$ it $\vee\ u\ E\ y)$]
being an entailer of y

that are not covered by items (1) to (5) of 1.A4.1 (quoted below). Being an entailer of y will, in certain cases, qualify as an attribute by being exemplified. But in SRD there is room for unexemplified attributes (null attributes, 1.5.5D) and, except for primitive attributes, being exemplified is not a required condition. In particular, nothing entails being a null attribute, but being an entailer of being identical with nothing, (i.e. of the empty set), which is a null attribute should be a null attribute.

The intent, as hinted at earlier, is to certify a putative attribute if it is intelligible (related in an approved way to certified, ultimately to exemplified, attributes) regardless of exemplars. Recursive treatment of primitive attributes and their Boolean compounds works towards that aim. Given, in a case of 1.5.6D, that x for 'it' is *predle*, so y is *predle*, so y is an *att*, and open quantification, as a primitive, is intelligible, then clearly it is intelligible that $(u)(\sim u\ E\ x \vee u\ E\ y)$, so the RHS of 1.5.6D should qualify as a referential phrase for an attribute. But it remains to specify the approved connection with other attributes that provides the formal warrant.

What has just been declared intelligible? That $(u)(\sim u\ E$ it $\vee\ u\ E\ y)$ (given the required conditions). Here I am dealing with a *predication* (something sayable), not an attribute. That predication is involved in the attempted specification of an attribute, in 1.5.6D. Since an abstract '[ϕ(it)]' is open to any determination there is no limit to the sentential complexity that must be handled. Therefore the final step (6) in 1.A4.1 must admit any *interpreted* abstract, whose notational content, with a suitable subject, expresses a predication.

Thus the chain of intelligibility starts in a user's subject matter where primitive attributes are exemplified and, as already shown, carries on through Boolean compounds of attributes. The present task is to specify predications on the basis of attributes already certified by 1.A4.1. That done, the chain of intelligibility goes back to attributes via a supplementary antecedent (6) for 1.A4.1, which is, in full, as follows.

1.A4.1 (1) $y\ E\ B f\ \&\ D\ x = B f$
(2) $\vee\ B\ r\ u\ Z\ \&\ B = y\ Z\ \&\ D\ x = B\ r\ y$
(3) $\vee\ B f Z\ \&\ D\ x = B\ n f$
(4) $\vee\ B f Z\ \&\ B\ g\ Z\ \&\ D\ x = B f\ .\ g$
(5) $\vee\ B f Z\ \&\ B\ g\ Z\ \&\ D\ x = B f : g$

$$(6)\ v\ \Theta\ \phi(v) \text{ is a } \textit{predcn}\ \&\ D\ x = [\phi(\text{it})] \supset x \text{ is an } \textit{att}$$

'*D*' for 'dependently', governs a formula, so its scope includes all that follows it in the foregoing lines. It is a theorem that $D\,x = x$ (though not that $x = x$), so by (6),

1.1 $\Theta\ \phi(v)$ is a *predcn* $\supset$ $[\phi(\text{it})]$ is an *att*

What, then, is a predication, and how does it relate to a proposition?

Kinds of discourse

A large amount of discourse, for some people the bulk, does not involve truth or falsity. This includes novels, movies, TV shows, fairy tales, comic strips, operas, poetry, much philosophy, pure mathematics, social-"science" discourse, 20th-century logical analyses of natural science, but not SRD, which is saddled with J. These make no, or an extremely tenuous, call on ascertainable public subject matter. They are quite literally fiction, made up without regard to any recalcitrant brute topics that might give the lie to them. Accordingly, a system of discourse with any claim to versatility should provide means for such portions of such discourse as are publicly intelligible. There should be a broader category of merely intelligible discourse within which a narrower category is concerned with the furniture of the world, whose condition determines truth or falsity. "Meaning is broader than truth."

In SRD, freedom of discourse is promoted by the employment of blankmarkers, and zulogy provides a foot on the ground of actuality. Ironically, contemporary logics typically require adoption of a "universe of discourse", which determines ranges of substituends for "variables", so non-topic-bound discourse is from the start impossible.

I here provide predications for the broad field of intelligible discourse, without regard to the relevance of truth and falsity. Fictive predications are independent of topic constraint — apparent subject matter depends on what is said. Propositions are dependent for their truth or falsity, but not their intelligibility, on non-made-up subject matter.

Aggregates

Booleans are Booleans of aggregates of items. *x* with *y* (x ‘ y) was called on in Volume 1 in treating Booleans of attributes, but the notation was not formally introduced. Volume 2 gave a comprehensive treatment of aggregates, so the gap has been filled. So aggregates will be called on freely in this volume.

PREDICATIONS

Specifications

A primitive notation 'Θ', a capital theta, is pronounced "that", though it is quite different from the 'that' of restrictive clauses and the demonstrative 'that'. When prefixed to a suitable notation (later to be taken as a schematic or a determinate sentence) it gives an expression referring to the predication that the sentence expresses. Using 'Θ', I now provide a recursive definition of predications. The seed of the recursion requires something intelligible as a description and some non-arbitrary expression for at least ostensibly referring to something so described. Whether or not there is such a referent is immaterial, so 'Oliver Twist' works as well as 'Aristotle'.

2.1D B a *predcn* = [

(1) $B = y$ is an *att* & [ϕ(it)] is an *att* & D(it = Θ $\phi(y)$)

(2) v Θ $\phi(y)$ is a *predcn* & D(it = Θ $(u)\phi(u)$)

(3) v Θ p is a *predcn* & D(it = Θ ~ p)

(4) v Θ p is a *predcn* & Θ q is a *predcn* & D(it = Θ p & q)]

2.1D covers the two primitive forms of wff, singular and openly quantified, and the two primitive kinds of compound wff, negative and conjunctive (not, as in PM, alternative). Note that no exemplar is called on. All that is needed is an identity attribute, $B = y$, for one. Such an identity attribute is automatically available for any zon, but in the absence of any suitable zon a suitable definition can be provided. Note that the embedded 'it' in (1) is not available for substitution, since any determination of '[ϕ(it)]' is a constant, though the top-level 'it', following 'D(', is.

2.1 $B = y$ is an *att* & [it is f] is an *att* $\supset$ Θ y is f is a *predcn*

2.2 Θ p is a *predcn* $\supset$ Θ ~ p is a *predcn*

2.3 Θ p is a *predcn* & Θ q is a *predcn* $\supset$ Θ p & q is a *predcn*

2.4 Θ $\phi(y)$ is a *predcn* $\supset$ Θ $(u)\phi(u)$ is a *predcn*

Certain definitions for predications, correlates of equivalences in sentential calculus and predicate calculus provide the beginnings of inference concerning predications.

2.2D Θ ~~p = Θ p

2.3D Θ q & p = Θ p & q

2.4D Θ (p v q) = Θ ~ (~ p & ~ q)

2.5D Θ $(\#u)\phi(u)$ = Θ ~(u)~ (u Z & $\phi(u)$)

The 'u Z' condition is not redundant. Since '$\phi(u)$' can be determined as '~ $\phi(u)$', it does not warrant that u Z. The zulogical clause has to be independently established. Conjunction is taken as primitive, and not alternation, because conjunction is called on in the very first axiom of SRD, and (covertly) in every theorem, as will be shown in Volume 6.

One further definition for attributes, not given in Volume 1, provides corresponding identities for attributes.

2.6D $B\ n\ n f = B f$
being not not f

Finally, I show that a proper closure of $B \Rrightarrow y$ is indeed an attribute.

a $B f$ is *predle* & $(u)(\sim u$ is $f \vee u\ E\ y) \supset y$ is *predle*
b $\supset B f$ is an *att* & y is an *att*
c $\supset (B = x$ is an *att* $\supset \Theta x$ is f is a *predcn* & $\Theta x\ E\ y$ is a *predcn*) (*b*, 2.1)
d $\supset \Theta$ J is f is a *predcn* & Θ J $E\ y$ is a *predcn* (B = J is an *att* 1.4.21)
e $\supset \Theta \sim$ J is $f \vee$ J $E\ y$ is a *predcn* (*d*, 2.2, 2.3, 2.4D)
f $\supset \Theta (u)(\sim u$ is $f \vee u\ E\ y)$ is a *predcn* (*e*, 2.4)
g $\supset \Theta B f$ is *predle* & $(u)(\sim u$ is $f \vee u\ E\ y)$ is a *predcn* (*b*, *f*, 2.3)
h $\supset$ [it is *predle* & $(u)(\sim u\ E$ it $\vee u\ E\ y)$] is an *att* (*g*, 1.1)
2.5 $\supset B \Rrightarrow y$ is an *att* (1.5.6D, above)

Predicational Booleans and components

There are Booleans of predications parallel to the Booleans of attributes treated in Volume 1.

2.7D B *P-neg* Θp = [Θp is a *predcn* & it = $\Theta \sim p$]
being a propositional negation of that p

2.6 $\Theta \sim p$ *P-neg* $\Theta p \equiv \Theta p$ is a *predcn*

2.8D B *P-conj* Θp ‘ Θq = [Θp is a *predcn* & Θq is a *predcn* & it = $\Theta (p \,\&\, q)$]
being a propositional conjunction of that p with that q

2.7 $\Theta p \,\&\, q$ *P-conj* Θp ‘ $\Theta q \equiv \Theta p$ is a *predcn* & Θq is a *predcn*

2.9D B *P-alt* Θp ‘ Θq = [Θp is a *predcn* & Θq is a *predcn* & it = $\Theta (p \vee q)$]
being a propositional alternation of that p with that q

2.8 $\Theta p \vee q$ *P-alt* Θp ‘ $\Theta q \equiv \Theta p$ is a *predcn* & Θq is a *predcn*

Note that a conjunction is a conjunction of an *aggregate* of items. Consequently, conjunctions and alternations are commutative and associative. I defer extending SRD to express this. Conditionals receive separate extended treatment in the next chapter.

Again, parallel to attributes, there are Boolean components of predications.

2.10D B *P-conjt* $\Theta p \,\&\, q$ = [Θp is a *predcn* & Θq is a *predcn* & it = $\Theta p \vee$ it = Θq]
being a propositional conjunct of that p and q

2.9 Θp *P-conjt* $\Theta p \,\&\, q \equiv \Theta p$ is a *predcn* & Θq is a *predcn*

2.10 Θq *P-conjt* $\Theta p \,\&\, q \equiv \Theta p$ is a *predcn* & Θq is a *predcn*

2.11D B *P-altt* $\Theta p \vee q$ = [Θp is a *predcn* & Θq is a *predcn* & it = $\Theta\ p \vee$ it = Θq]
being a propositional alternate of that p or q

2.11 Θp *P-altt* $\Theta p \vee q \equiv \Theta p$ is a *predcn* & Θq is a *predcn*

2.12 Θq *P-altt* $\Theta p \vee q \equiv \Theta p$ is a *predcn* & Θq is a *predcn*

Volume 3 provides an account of *elementary* components of Booleans.

Say-so

I adopt 'so' (instead of 'true') to cover fictional assertions as well as factual ones. The latter are true or else false but the former are neither. In a fictional sense, they are *so*, but they are accepted not by belief but, by suspension of disbelieve, on the author's *say-so*. I do not deny that Oliver Twist was an orphan but I do not affirm it, and I neither believe nor disbelieve it. Affirmation, denial, belief, disbelief, knowledge, truth and falsity are all reserved for propositions. Each fictional discourse and each axiomatic system claims a "world" of its own. It is pointless to dispute what an author of fiction says.

2.12D B so = [(it = Θp) & p]

2.13 $\Theta p Z \supset$ (Θp is so $\equiv p$)

2.14 Θp is a *predcn* $\supset$ (Θp is so $\equiv p$)

Open affirmability provides that *suitable determinations* of open formulas express so predications, the conditions being explicit. There is no open simple predication corresponding to an open simple formula; that would open the door to gibberish.

a	B = J Z & [it Z] is an *att*	(Volume 1)
b	Θ J Z is a *predcn*	(2.1)
c	J Z	(Volume 1)
d	Θ J Z is so	(2.14)
e	B so is *predle*	
2.15	B so is an *att*	
a	Θp is n so $\equiv \Theta p Z$ & B so is an *att* & ~ Θp is so	[definition in Volume 1]
2.16	Θp is n so $\equiv \Theta p Z$ & ~ p	(2.15, 2,13)
2.17	~ $p \supset$ (Θp is n so $\equiv \Theta p Z$)	
2.18	$\Theta p Z \supset$ (Θp is n so $\equiv$ ~ p)	

As usual in SRD, the definition of negation prevents purported assertion when there is nothing to assert.

Linking identity and affirmability

The definitions for predications permit (intensional, not textual) metadiscussion, making reference to predications; such discussion is independent of any assertion of the predications. But a link from these identities to assertion is available, so that meta-inferences can have corresponding plain inferences.

2.19 $\Theta p = \Theta q \supset$ (Θp is so $\equiv \Theta q$ is so)

2.20 $\Theta p = \Theta q \supset (p \equiv q)$ (2.19, 2.13)

Clearly, an identity for predications provides a logical equivalence for them. But the converse is not forthcoming, and is not desirable. Equivalence is weaker than identity, but is all that is needed for dealing with what follows from what. The original equivalences behind 2.2D – 2.5D were elevated to identities because they genuinely deal with meanings in the first place. But equivalence as such has nothing to do with meaning, so standard logic, lacking any way to refer to predications or attributes, is locked out of the realm of meaning. That is a cause of many of its problems, which do not get solved but just lead to further problems.

Note the components of the identity and those of the equivalence in 2.20. In the former, the component predications are *mentioned.* In the latter, they are *expressed.* In that context, '$p = q$' or '$\Theta\, p \equiv \Theta\, q$' would, like '$p$ implies q', be incoherent. On this basis, the whole of verbalized mathematics is incoherent. Mathematicians and natural scientists don't care, and *as such,* they do not need to, but when they venture into philosophy or popular exposition they are out of their depth — and it shows.

By 2.13 and 2.18,

a $\Theta\, p$ is a *predcn* $\supset (\Theta\, p$ is so $\vee\, \Theta\, p$ is n so $\equiv p \vee {\sim} p)$

b $\supset \Theta\, p$ is so : n so

c $\Theta\, p$ is a *predcn* $\supset (\Theta\, p$ is so & $\Theta\, p$ is n so $\equiv p\, \&\, {\sim} p)$

d $\supset \Theta\, p$ is n (so . n so)

2.21 $\Theta\, p$ is a *predcn* $\supset \Theta\, p$ is (so : n so) . n (so . n so) (*b*, *d*)

In SRD, any predication is either so or else not so.

The sooth value of a compound is determined by the values of its components.

a $\Theta\, p$ is a *predcn* $\supset \Theta\, {\sim} p$ is a *predcn*
& $(\Theta\, {\sim} p$ is so $\equiv {\sim} p)$ & $(\Theta\, p$ is n so $\equiv {\sim} p)$

2.22 $\Theta\, p$ is a *predcn* $\supset (\Theta\, p$ is n so $\equiv \Theta\, {\sim} p$ is so)

2.23 $\Theta\, p$ is a *predcn* $\supset (\Theta\, p$ is so $\equiv \Theta\, {\sim} p$ is n so)

2.24 $\Theta\, p$ is a *predcn* & $\Theta\, q$ is a *predcn*
$\supset (\Theta\, p\, \&\, q$ is so $\equiv p\, \&\, q)$

2.25 $\supset (\Theta\, p\, \&\, q$ is so $\equiv \Theta\, p$ is so & $\Theta\, q$ is so)

2.26 $\supset (\Theta\, p \vee q$ is so $\equiv \Theta\, p$ is so $\vee\, \Theta\, q$ is so)

2.27 $\supset (\Theta\, p\, \&\, q$ is n so $\equiv \Theta\, p$ is n so $\vee\, \Theta\, q$ is n so)

2.28 $\supset (\Theta\, p \vee q$ is n so $\equiv \Theta\, p$ is n so & $\Theta\, q$ is n so)

SRD is thus *sooth*-functional, becoming *truth*-functional when restricted to propositions. Besides sooth/nonsooth and truth/falsity (see below) there is a further dichotomy, that between theorems and their negations. So SRD is a two-valued calculus in a variety of ways, as explained at length in Volume 6. Again one sees that SRD has *room to move* so can find straightforward solutions to problems that the too restricted basis of standard logic labours with unsuccessfully. For the same reason, SRD can incorporate many standard calculi, showing their likenesses and differences, and usually their deficiencies compared with SRD proper.

CONDITIONALS

Here I just say what conditionals and their components are in SRD. The complications that are standard topics of controversy are dealt with under implication and under vernacular quantifiers.

In a conjunction or an alternation, call is usually made on multiple predications; the compound is a compound of (some aggregate of) components. The commutativity of aggregation fits well with conjunction and alternation but it does not suit a conditional, which is polarized. The basic expression for a conditional, being, simply or with some complication, '$\sim p \vee q$', distinguishes the components by the single negation, and that distinction carries over to polarize the conditional expression '$p \supset q$', where the negation is implicit, being bound up in the conditional connective. Whereas $\Theta \sim p \vee q$ is identical with $\Theta\, q \vee \sim p$, $\Theta\, p \supset q$ is not even generally equivalent to $\Theta\, q \supset p$.

An aggregate is, as such, not polarized. Consider an aggregate of predications $\Theta\, q$ with $\Theta\, p$ with $\Theta\, r$. These are a few of the conditionals of it: $\Theta\, p \supset q\ \&\ r$, $\Theta\, r \supset q\ \&\ p$, $\Theta\, p \vee q \supset r$. Since any one item (out of 3) can be chosen as antecedent (subtotal 3), and each pair may be put into conjunction or alternation (× 2), and each conditional can have its antecent and its consequent interchanged (× 2), there are 12 conditionals of the aggregate. There is no such thing as *the* conditional of any aggregate of predications. Resort to "ordered n-tuples" has been dismissed in Volume 2.

A conditional, unlike a conjunction or an alternation, is polarized, so the aggregate called on must accord with the polarization, by having relevantly distinguishable members. Plain blankmarkers do not make any distinction so a special step is needed. This involves defining the components of a conditional before the compound.

3.1D *B ante* $\Theta\, (p \supset q)$ = [$\Theta\, p$ is a *predcn* & $\Theta\, q$ is a *predcn* & it = $\Theta\, p$]
being an antecedent of that if *p* then *q*

3.1 $\Theta\, p$ *ante* $\Theta\, (p \supset q)$ ≡ $\Theta\, p$ is a *predcn* & $\Theta\, q$ is a *predcn*

3.2D *B cons* $\Theta\, (p \supset q)$ = [$\Theta\, p$ is a *predcn* & $\Theta\, q$ is a *predcn* & it = $\Theta\, q$]
being a consequent of that if *p* then *q*

3.2 $\Theta\, q$ *cons* $\Theta\, (p \supset q)$ ≡ $\Theta\, p$ is a *predcn* & $\Theta\, q$ is a *predcn*

3.3 $\Theta\, p$ *ante* $\Theta\, (p \supset q)$ ≡ $\Theta\, q$ *cons* $\Theta\, (p \supset q)$ (3,1, 3.2)

Here the notation '$p \supset q$' itself provides the polarization, but '*p* ' *q*' does not, therefore the object of the relation must be more fully specified.

3.3D *B condl* $\Theta\, p$ *ante* ' $\Theta\, q$ *cons* =
[$\Theta\, p$ is a *predcn* & $\Theta\, q$ is a *predcn* & it = $\Theta\, p \supset q$]
being a conditional of that *p* as antecedent with that *q* as consequent

3.4 $\Theta\, p \supset q$ *condl* $\Theta\, p$ *ante* ' $\Theta\, q$ *cons* ≡ $\Theta\, p$ is a *predcn* & $\Theta\, q$ is a *predcn*

Commutation of the qualified members of the aggregate does not affect the outcome.

3.5 $\Theta\, p \supset q$ *condl* $\Theta\, q$ *cons* ' $\Theta\, p$ *ante* ≡ $\Theta\, p$ is a *predcn* & $\Theta\, q$ is a *predcn*

Definition of a biconditional does not require polarized components.

3.4D *B bicon* $\Theta\, p$ ' $\Theta\, q$ = [$\Theta\, p$ is a *predcn* & $\Theta\, q$ is a *predcn* & it = $\Theta\, p \equiv q$]

3.6 $\Theta\, p \equiv q$ *bicon* $\Theta\, p$ ‘ $\Theta\, q$ $\equiv$ $\Theta\, p$ is a *predcn* & $\Theta\, q$ is a *predcn*

Conditionality holds between predications regardless of their character, as shown by the blankmarkers ‘*p*’, etc.. No referential blankmarkers or predicate blankmarkers have been called on in this chapter. So the force of conditionality is independent of the character of involved propositions; it comes from some *mode of affirmation*. A conditional cannot be a simple affirmation. Generally, conditionals are *theorems* in a system, not assertions about external subject matter.

“IMPLICATION” AND IMPLICATION

“Implication”

It is doubtful whether any formula of a coherent formal system has ever yet been uttered that contains any reasonable analogue of ‘implies’. Occurrences of ‘implies’ float around in a vernacular commentary on, perhaps never a translation or encoding of, formulas that are suitably, if at all, rendered as ‘if *p* then *q*’. By contrast, this present calculus has from the start used a formula ‘$p \supset q$’ as a written abbreviation, to be spoken in ordinary English as “if *p* then *q*”, without the remotest suggestion that it could be intelligibly spoken as “*p* implies *q*”. Authors may, if they choose, claim that a formal system does coherently treat ‘*p* implies *q*’; if so, what corresponds to ‘if *p* then *q*’? No formula can do both in a coherent system.

There are two reasons. One is Quine’s bugbear, the confusing of use and mention, a fault, imported from PM, that is rampant in discussions of modal logic (indeed just rampant), a fault that he confesses he could never explain to certain of his eminent colleagues. They could not be got to see that they were uttering gibberish. That mathematicians institutionally do so with ‘*p* implies *q*’ just proves that their work has no dependence on English renderings and no import for English renderings, with consequences that have been shown at length in previous volumes. I have found the same difficulty as Quine. What adds to intellectual misfortune is that even careful workers like Quine and Tarski follow the rules of rigorous but *inappropriate* methods of treating text. The fault and the correct method (which everybody informally uses as part of the vernacular when not trying to be precise) will be treated in Volume 5.

There are some who claim to be aware of the distinction between use and mention but say that it is pedantic and that disregard of it can be harmless. They thereby flout the fundamental principle of Logic: that there are *no* unimportant distinctions for logicians, whatever corners *users* of logic may be permitted to cut. That is the reason for SRD’s maxim: *if a hair can be split, split it.*

It is in part to avoid the use-mention confusion altogether that I develop SRD without call on notation at all. If notations are never mentioned (outside of informal expository discussion), even the most obdurate confusers will (one can only hope!) find no opportunity to exercise their flair. But the outlook for reform is bleak. Dissuading the obstinate may be harder than educating the ignorant.

The other reason is that conditionality, in any form, is not implication. I here introduce formulas for various cases where ‘implies’ has been imported, in uses warranting raised-eyebrow quotes, which are presumed throughout this section. I dignify these cases with properly numbered definitions and distinctive notations because they are explicable in SRD even if they have no useful applications, and SRD includes the systems containing them. But they are as non-functional in SRD as the exhaust noise of a car is in driving.

4.1D *B Ml-impl y* = [it is a *predcn* & *y* is a *predcn* & (~ it is so ∨ *y* is so)]
being a material implier of *y*

As is customary in SRD, the new predicate notation may be read "materially implies".

4.1 $\Theta\, p$ *Ml-impl* $\Theta\, q \equiv \Theta\, p$ is a *predcn* & $\Theta\, q$ is a *predcn* & $(\sim p \vee q)$

4.2 $\Theta\, p$ is a *predcn* & $\Theta\, q$ is a *predcn* $\supset$ ($\Theta\, p$ *Ml-impl* $\Theta\, q \equiv \sim p \vee q$)

Material implication is extremely unselective. A not so predication materially implies any predication, and any predication materially implies a so predication.

4.3 $\Theta\, p$ is *n* so & $\Theta\, q$ is a *predcn* $\supset$ $\Theta\, p$ *Ml-impl* $\Theta\, q$

4.4 $\Theta\, p$ is a *predcn* & $\Theta\, q$ is so $\supset$ $\Theta\, p$ *Ml-impl* $\Theta\, q$

Strict implication, associated with a necessary conditional, is restricted to theorems. Being a strict implier of *y* cannot be defined in SRD since it would require a modal formula inside an abstract, which is illicit. This is not an arbitrary proscription. An abstract is concerned with meaning, not with assertion, hence not with a necessary mode of assertion. Volume 6 will make this abundantly clear. So I present here the standard way of "defining", i.e., making an axiomatic equivalence. The asterisk indicates that the axiom is not to be taken seriously.

***4.2D** $\Theta\, p$ *St-impl* $\Theta\, q \equiv \Theta\, p$ is a *predcn* & $\Theta\, q$ is a *predcn* & $L\,(\sim p \vee q)$

4.5 $\Theta\, p$ is a *predcn* & $\Theta\, q$ is a *predcn* $\supset$ ($\Theta\, p$ *St-impl* $\Theta\, q \equiv L\,(\sim p \vee q)$

It is well known that strict implication has its paradoxes, parallel to those of material implication. In both cases, the vacuous cases quite clearly state what follows from what — they are theorems about deducibility among unanalysed predications whose subject matter is not constrained. Seen as such they are, quite obviously, not paradoxes. But 'if *p* then *q*' has for long been discussed under the rubric 'implication'. It does seem peculiar, on an ordinary understanding of implication, that, e.g Θ snow is blue implies Θ grass is red, and Θ snow is red implies Θ grass is green, since the premise is not relevant to the conclusion and one feels that implication requires relevance. So these vacuous cases are not cases of implication; *a fortiori*, they are not paradoxes of implication. The mere suggestion that they are paradoxes stems from a false appreciation of them, leading to misleading terminology that generates unfounded expectations. But usage is too entrenched to be disregarded here.

Another standard case makes provision for relevance by employing a common referential expression for antecedent and consequent, thus preventing change of subject matter and attaining relevance.

4.3D *B Fl-impl y* = [*B f* is an *att* & *B g* is an *att*
& $(u)(\sim u$ is $f \vee u$ is $g)$ & it = $\Theta\, x$ is f & $y = \Theta\, x$ is g]
being a formal implier of *y*

4.6 *B f* is an *att* & *B g* is an *att* $\supset$
($\Theta\, x$ is *f Fl-impl* $\Theta\, x$ is $g \equiv (u)(\sim u$ is $f \vee u$ is $g)$

Being dependent on material implication, this case loses relevance in vacuous cases, e.g. that Russell is a unicorn formally implies that Russell lives in the Bronx Zoo, (along with "all the other unicorns"). If strict implication is called on, similar conclusions emerge.

Implication

As already said, many difficulties stem from confusing conditionality and implication. The force of conditionality is attained in SRD by means to be described in Volume 6. Lewis tackled that matter in his theory of strict implication, but confusions remain in the whole framework of modal logic. But meaningful content and relevance are called on for implication, and these are unaffected by the misnamed strict "implication". Relevance was attained in formal "implication", (note that it calls on predicates), but relevance fails in vacuous cases. Given a sound treatment of conditionality (Volume 6), the remaining problem is that of vacuous cases, and that has already been solve in SRD's treatment of entailment, which excludes them. So implication proper, simpliciter, is easily defined, in a manner that avoids all the standard problems.

4.4D *B impl y* = [[ϕ(it)] ⇛ [φ(it)]] & *D* it = Θ ϕ(*x*) & *D y* = Θ φ(*x*)]
being an implier of *y*

Here, again, I show a case of abstracts embedded in an abstract. According to the rule in Volume 1, the embedded 'it's, acting like standard "bound variables" are immune when substitution is made for the non-embedded 'it'. The rule would be unnecessary if recourse were made to multiple pronouns.

4.7 Θ ϕ(*x*) *impl* Θ φ(*x*) ≡ [ϕ(it)] ⇛ [φ(it)]

4.8 Θ ϕ(*x*) *impl* Θ φ(*x*) ≡
[ϕ(it)] is *predle* & [φ(it)] is an *att* & (*u*)(~ ϕ(*u*) v φ(*u*))

Any attribute entails an attribute just if the first-mentioned is predicable and every exemplar of the first-mentioned is an exemplar of the second-mentioned. The associated predications are then related in implication.

As with entailment, co-implication and proper implication follow routinely.

4.**5D***B co-impl y* = [it *impl y* & *y impl* it]
being a co-implier of *y*

4.9 Θ ϕ(*x*) *co-impl* Θ φ(*x*) ≡ [ϕ(it)] is *predle* & (*u*)(ϕ(*u*) ≡ φ(*u*))

4.6D*B pro-impl y* = [it *impl y* & ~ *y impl* it]
being a proper implier of *y*

4.10 Θ ϕ(it) *pro-impl* Θ φ(it) ≡
[ϕ(it)] is *predle* & (*u*)(~ ϕ(*u*) v φ(*u*)) & (#*v*)(~ ϕ(*v*) & φ(*v*))

Note that no call is made here on necessity. Necessary implication requires necessary assertions. Related matters, including standard pseudo-paradoxes, are treated below in the context of vernacular quantifiers.

CLOSER TOWARDS THE VERNACULAR

This chapter provides further evidence, on top of that given in Volume 1, that the English language (like, I do not doubt, many others) is a superb means of rational expression. Its versatility and terseness (never in doubt) are (common criticisms notwithstanding) compatible with full precision, when its expressions are suitably used. One cannot say "properly used" or "correctly used" without invoking some (surely controversial) evaluative criterion. But (unofficial) usage that gives rise to paradox and (official) usage that gives rise to gibberish in modern logic are surely unsuitable.

Existence and thinghood

Some people claim that *some thing* does not exist. Does this claim imply that *something* does not exist? Does this imply that *something* is not a thing (easily defensible against a charge of inconsistency)? Does this imply that *some thing* is not a thing (much harder to defend)? Do these problems arise from associating 'thing' as an independent word too strongly with '...thing' as merely a part of a word? Is doing this "treating existence as a predicate"? And is such treatment wrong?

These questions undoubtedly bear on important issues. But what are they? Philosophers who apparently argue these issues, in whatever language, might as well be a monoglot Englishman, a monoglot Chinese, a monoglot Martian, etc.. Perhaps they are all correct, but none of them can demonstrate that. It is easy to see how metaphysics got a bad name, but it is a pity that natural language got the blame for the troubles caused by various philosophers' varying usage.

Usage in SRD is disciplined. Is [it *Z*] an attribute? Yes. Is '*Z*' a predicate notation, usable after 'is'? No; as the voiced sibilant is intended to suggest, '*Z*' replaces and disciplines the non-copulative 'is'. Can a corresponding predicate notation be introduced? Yes. Is it useful to do so? Yes. Then what shall it be? The answer is given by 2.1.11D, introducing a *zon*. Its character is entirely given by the definition but the suggestion of rigorous **z**ulogy and familiar **on**tology does no harm. From the definition,

5.1 x is a *zon* $\equiv x\,Z \equiv x = x \equiv (\#u)(u = x)$

As noted in Volume 1, '$(\exists u)(u = x)$' has no zulogical bearing. To say that '$\exists$' is synonymous with 'there exists' carries about as much weight as saying that '$\supset$' is synonymous with 'implies'. Saying this, even loudly by means of an axiom, may make it so, in some Pickwickian usage in some system, but does not make it true for English.

The ontological argument

Many traditional arguments boil down to the following: "Existence is part of the definition of a perfect being, therefore there is a perfect being.". Consider any proposed definition for *B* a perf bg.

D *B* a perf bg $=$ [ϕ(it) & it *Z*]

a x is a perf bg $\equiv x\,Z\ \&\ \phi(x)\ \&\ x\,Z$

The first '$x\,Z$' comes from the axiom for abstract introduction, the second from the 'it *Z*' in the abstract; the latter is redundant, and does not provide the desired conclusion.

It must be otherwise established that *x Z* if the defined attribute is to find an exemplar. That principle is the first one to be laid down in SRD, and has been put to use over and over again, so it cannot be truly said that it was made up just for the sake of the present example.

According to one sort of corollary of the previous argument: "Because it adds nothing to the description of an object to say that it exists, therefore existence is not a predicate". By contrast, for SRD, because one *can* add to a definiens a conjunct that it Z, *zon*hood is intelligible and an attribute, and its expression takes the standard reference-forming suffix for a common noun. Still, the addition makes no difference regarding exemplars, so 'Z' is not idle.

Not only can being a *zon* be defined. Being a *nonzon* can also.

5.1D *B* a *nonzon* = [~ it Z]

5.2 *x* is a *nonzon* ≡ *x* Z & ~ *x* Z

5.3 (*x*)~ *x* is a *nonzon*

As in the previous example, the outcome is not changed by adding a condition that *x* Z. That merely makes the definiens immediately rather than mediately inconsistent. Other inconsistent definitions have already been met in SRD, and they cause no trouble whatsoever. What can be harmful with an inconsistent meaning? The case is different for standard logics. There, it is not attributes that are defined; rather, notations or exemplars are "defined" (the latter in effect created by axiom), and lead to many troubles. Instead of the troubles' being viewed as a cause for suspicion about the value of the system, they are in many cases viewed as "genuine paradoxes" *inherent in the intellectual enterprise*. I judge it a sign of wavering intellectual resolution that such a stand can be even contemplated.

a *B* a *zon* θ *f* = *B* a *zon* . *f* (1.6.2D)

5.4 *x* is a *zon* . *f* ≡ *x* is *f*

The relevance of all this to the ontological argument is incidental. The intended purpose of '*zon*' is for the introduction of vernacular infix quantifiers.

5.5 *x* is a *zon* θ *a b* ≡ *x* is a *zon* . *a b* ≡ *x* is a *b*

'A *zon* θ' is redundant in assertions. But use of '*zon*' provides noun phrases from adjectives, leading to a uniform treatment of superficially dissimilar predications.

Eliminating Fregean quantifiers

It is to Frege's credit, as acknowledged by some, that he introduced prefix quantifiers, thereby providing, as I show here, a solid basis for clearing up many traditional confusions and controversies. But, as I have shown repeatedly, the new quantifiers introduced troubles of their own, brought on by their vacuous cases, so that standard logic is not able to confirm all the details of traditional syllogistic, and puts the blame on tradition. It is not to his credit that his mathematical interest and skill so blinkered him that he failed to apply his invention to natural language and import its rigour. His successors, instead of doing that, have acted as if that invention were ideal in itself.

But in fact the useful modes of inference are those engendered (up to now without rigour) by vernacular quantifiers, and the new rigorous *but undeveloped* modes of inference introduced by Frege's invention are not logically useful. Since the avowed aim of SRD is to achieve both rigour and useful inference, it is an obvious step to use the rigour of prefix quantifiers to make use of infix quantifiers precise. To this I turn.

There are two steps. The goal is to treat the infix quantifiers 'any', 'every', 'no' and 'some', but I first show that the work of prefix quantifiers can also be duplicated in English without any strain on either expression or meaning. For *open* (usually called universal) quantifiers, I introduce 'wti' to abbreviate 'whatever there is, is ...'.

5.2D Θ wti f = Θ $(u)(\sim u\ Z \vee u$ is $f)$
that whatever there is, is f

Use of 'wti', like use of '(u)', does not require that anything zu, or even a genuine predicate; it is open and *completely non-committal.* 'Tia' is used to abbreviate 'there is a ...'; this is a *zulogical* quantifier; here a genuine predicate is guaranteed through predicability.

5.3D Θ tia b = Θ $(\#u)(u$ is a $b)$
that there is a b

Logical equivalences follow from these definitions, by 2.19.

5.6 wti $f \equiv (u)(\sim u\ Z \vee u$ is $f)$

5.7 tia $b \equiv (\#u)(u$ is a $b) \equiv \sim (u)\sim (u\ Z\ \&\ u$ is $a\ b)$ (see 1.A1.3)

There are two points of interest in comparing zulogical quantification and "existential" quantification. The first, pointed out in Volume 1, is that the latter, being wholly contained in a negation, without a zulogical clause, is a dead end; nothing follows from a pure negation; some affirmative contribution is needed for inference. The second is the common use of 'there is (exists) a'. SRD's 'there is a *b*' takes a common noun, and is grammatical in English but the standard 'there exists an *x* such that', which takes a referential expression following an article, is ungrammatical gibberish, with dire results, as shown next.

The logical weakness of sentential quantifiers

Sentential quantifiers, whether Fregean or the corresponding English ones just introduced, are perfectly sound, and a necessary first step in the development of a useful system of logic. But that is all they are: a first step, not the last word. One weakness is that vacuous cases render theorems unsuitable for serious inference. Others are the lack of zulogical quantification and the absence of polarization, due to *the lack of common nouns*. No component is distinctively related to any other or to the whole.

5.8 wti $n\,f : g \equiv$ wti $g : n\,f$

5.9 tia *zon* $\theta\,f\,.\,g \equiv$ tia *zon* $\theta\,g\,.\,f$

5.10 wti $n\,f \supset$ wti $n\,f : g$

a wti *n* a unicorn $\supset$ wti *n* a unicorn : a resident of the Bronx Zoo

b wti *n* a unicorn

c wti *n* a unicorn : a resident of the Bronx Zoo

This is a perfectly valid, innocuous but *useless* inference with a true conclusion: whatever there is, is not a unicorn, so whatever there is, is not a unicorn or in the Bronx Zoo. But to paraphrase this as 'Every unicorn is in the Bronx Zoo' or 'All unicorns are in the Bronx Zoo', as is commonly done, is just too silly to be contemplated in SRD, as explained in the discussion of implication above. As shown below, it is well paraphrased, noncommittally as '*Any* unicorn is in the Bronx zoo'. I guarantee any unicorn to be in Bronx zoo — *until you find one*; then I shall reconsider.

Open (i.e. "universal"), zulogical and "existential" predications do not "direct attention" to any items of a particular kind. 5.8 is no more "about" not-*f* items than "about" *g* ones. 5.9 is no more "about" *f* items than "about" *g* ones. In particular

Θ wti not a raven or black

or, identically

Θ wti not not-black or not a raven

is no more "about" ravens than it is "about" not-black items, and there is nothing in either the abbreviated text or the full English to suggest otherwise. If this universal were taken as an empirical hypothesis then observation of a single not-black raven would refute it (so far so good), and observation of black ravens, of black non-ravens and of not-black non-ravens would support it. That is Hempel's paradox of confirmation. taken as such because the open predication *is* (for lack of any alternative) taken as a hypothesis.

One has a strong feeling that a hypothesis that every raven is black is definitely "about" ravens in a way that it is not "about" not-black things or non-ravens, as others have argued, so far unsuccessfully. Ravens as such are, and not-black things and non-ravens as such are not, relevant to it. Though there is nothing paradoxical about the open predication or the mode of its confirmation, if it were (à la Hempel) taken as a raven hypothesis it would engender paradox. *Obviously*, it should not be so taken; that is the *lesson* of the apparent paradox.

Vernacular quantifiers

If someone claims that some horse is a unicorn we can ask for his evidence. Since his claim is "about" horses we shall expect to examine horses. If someone claims that some unicorn is a horse our reaction will be quite different. Factually: since there are no unicorns he has no subject matter. Fictionally: we can perhaps ask for his literary sources. Logically: disregarding the matter of what there is, his claim is analytic — fictively so by virtue of meaning. Clearly, that some horse is a unicorn is (in zoological vein) quite a different kettle of fish from that some unicorn is a horse. Both are intelligible (are predications) but one is a contingent proposition and is true or else false, and the other is an analytic non-proposition.

The foregoing is a presentation of some commonplace philosophical doctrine, often presented in a more longwinded and less disciplined manner. Non-technical philosophers are right in principle. Technical philosophers, particularly Quine, dismiss the usual presentations of doctrine as confused, ontologically extravagant and generally unworkable. They too are right, but, as too often, their criticisms destroy the presentations, not the doctrine.

Clearly, the doctrine requires that a 'some' predication, like its cognates, be polarized in some way. The indication of the claimed description (being a unicorn) must be distinguishable from the indication of the assumed character (being a horse) of the things described. Thus an "existentially" quantified formula, with its commutative conjunction, is not an adequate formulation, even if its other defects are disregarded. The clue lies in the presupposition, (maintained but not made explicit, and incapable of being expressed in LPC, which lacks zulogy) that there is some relevant subject matter of discussion. That presupposition, made explicit in a definiens, provides the needed polarization. A zulogical quantifier makes such a claim, but not in a way to distinguish the presupposed subject matter from the added comment.

Four vernacular quantifiers, 'any', 'every', 'no', and 'some', are initially called on, abbreviated to initial capital script characters. All but the first have zulogical import for the referential term and a meaningful expression for the predicate term; this condition captures the intuitive inferences on which traditional syllogistic is based, and explains why standard logics cannot reproduce the theorems of syllogistic.

5.4D $\Theta\ \phi(\mathcal{A}\ b)\ =\ \Theta$ wti n a b : E [ϕ(it)] & B a b is an *att* & [ϕ(it)] is an *att*
that any b E [ϕ(it)] (that ϕ(any b))

Though there is no zulogical import and there are vacuous cases, there is meaning import. SRD does not condone any utterance of meaningless expressions outside negations. To do such would be pointless. This definition and its cognates introduce quantified noun phrases as referential phrases (referential in function though dependent on subject matter for successful exercise of that function). They are therefore proper substituends for blank-markers in referential position. It is a test of the robustness of SRD that it tolerates such substitution without incoherence or gibberish. There is no possible correlate of such substitution in LPC, where the quantificational apparatus consists of a prefix disconnected from the "bound variable" that it binds.

5.5D $\Theta\ \phi(\mathcal{E}\ b)\ =\ \Theta\ \phi(\mathcal{A}\ b)$ & B a b is *predle*
that every b E [ϕ(it)]

Here, there is zulogical import for the referential expression. Some other quantifiers are defined in terms of this one. LPC has corresponding "definitions" in terms of open quantification and equivalence, or notational "definitions" and none of its quantifiers carries zulogical import, nor meaning import except by pre-systematic restriction of "variables" to certain ranges.

5.6D $\Theta\ \phi(\mathcal{N}\ b)\ =\ \Theta\ \mathcal{E}\ b\ n\ E$ [ϕ(it)]
that no b E [ϕ(it)]

Here the general clause carries the needed import but in the next case the import is lost in the negation, so must be reaffirmed.

5.7D Θ φ(ꝭ *b*) = Θ ~ Ɛ *b n E* [φ(it)] & *B* a *b* is *predle* & [φ(it)] is an *att* that some *b E* [φ(it)]

Nothing here calls on aggregates. Whatever is said concerns individual *b*s. Aggregates of *b*s, have, in general, different attributes from those of their members, as dealt with in Volume 2, although some features of quantification (not involving 'any') can be paraphrased in talk of inclusion among aggregates.

With *B f* for [φ(it)],

5.11 Θ Ɛ *b* is *f* = Θ wti *n* a *b* : *f* & *B* a *b* is *predle* & *B f* is an *att* that every *b* is *f* (5.5D)

There are corresponding results for the other quantifiers.

Distinctness, and even lack of unconditional equivalence, of the two particular predications (one "about" horses, the other purportedly "about" unicorns) is shown here.

5.12 Θ ꝭ *h* is a *uc* = Θ ~ Ɛ *h* is *n* a *uc* & *B* a *h* is *predle* & *B* a *uc* is an *att*

This correctly presupposes that there is a horse.

5.13 Θ ꝭ *uc* is a *h* = Θ ~ Ɛ *uc* is *n* a *h* & *B* a *uc* is *predle* & *B* a *h* is an *att*

This incorrectly presupposes that there is a unicorn, but is no more and no less intelligible than the other. The distinctness of the two predications is not one of meaning but one of reference. The one is "about" something and is false, the other is not about anything, and is not so. But that difference is irrelevant here. The non-equivalence is more stark with the negated predicate. Θ ꝭ *h* is *n* a *uc* is true and Θ ꝭ *uc* is *n* a *h* is not so.

A polarized predication that every raven is black, distinct from the predication that every not-black thing is a non-raven, serves well as an empirical hypothesis

5.14 Θ Ɛ raven is black = Θ (*u*)(*u* is *n* a raven v *u* is black) & *B* a raven is *predle* & *B n* black is an *att*

5.15 Θ Ɛ *zon n* black is *n* a raven = Θ (*u*)(*u* is *n* a raven v *u* is black) & *B n* black is *predle* & *B* a raven is an *att*

Though all the attributes called on here are indeed predicable, that is not relevant. What matters is that the identities call on different conditions so deal with different predications. The predications, though distinct, are, however, conditionally equivalent. At no time is affirmability relevant here. The concern is with what *can be* affirmed or denied or treated in any other way by use of the sentences, i.e. with their meanings.

Since Θ φ(Ɛ *b*) is neither affirmed nor denied in 5.5D, the following, which would be legitimate for equivalence, is unwarranted for identity.

* *B* a raven is *predle* & *B n* black is *predle* ⊃ (Θ Ɛ raven is black = Θ Ɛ *zon n* black is *n* a raven) (5.14, 5.15)

The antecedent here has been untimely ripped from a referential phrase, in the scope of 'Θ', not from a freestanding formula. But the corresponding equivalence is forthcoming.

a *B* a raven is *predle* & *B n* black is *predle* ⊃
(ℰ raven is black ≡ ℰ *zon n* black is *n* a raven) (5.15, 2.19)

Since the condition is fulfilled,

5.16 ℰ raven is black ≡ ℰ *zon n* black is *n* a raven

A result like this, a commonplace of ordinary English and unformalized philosophy, shows the poverty of standard logic, where there is no stronger relevant relation than equivalence.

The obvious results for quantifiers are forthcoming.

5.17 ℰ *b* is *f* ⊃ 𝒜 *b* is *f*

5.18 ℰ *b* is *f* & *x* is a *b* ⊃ *x* is *f*

5.19 *x* ω is a *b* is *f* ⊃ 𝒜 *b* is *f*

5.20 ℰ *b* is a *c* & ℰ *c* is *f* ⊃ ℰ *b* is *f*

5.21 𝒜 *b* is a *c* & ℰ *c* is *f* ⊃ 𝒜 *b* is *f*

There have never been any *obvious* results for 'no' — it has always been controversial. But in SRD, where every vernacular quantifier except '𝒜' has zulogical import,

5.22 𝒩 *b* is *f* ≡ ℰ *b* is *n f*

5.23 𝒩 *b* is *f* ⊃ 𝒜 *b* is *n f*

A 'no' assertion is not automatically convertible. Regarding horses,

5.24 𝒩 *h* is a *uc*

5.25 ~𝒩 *uc* is a *h*

Since there are no unicorns, the negation sign cannot be eliminated.

Quantification in relations

The foregoing treatment is perfectly general for the subjects of sentences, regardless of the number of noun phrases called on in relations. But English has further quantifiers for the objects of relations, particularly in negations. With '*B r y*' for '[ϕ(it)]', for a quantified subject,

5.26 Θ ℰ *b r y* = Θ 𝒜 *b r y* & *B* a *b* is *predle* & *B r y* is an *att* (5.5D)
that every *b* is *r* of *y*

This is not the place for a comprehensive treatment of relations but, in anticipation of that, the converse relation is introduced. Being conversely a parent is being a child.

5.8D *B* γ *r y* = [*y r* it]
being conversely *r* of *y*

An axiom is needed here to confirm the converse as an attribute but, in a full treatment of relations, a generalization of (2) in 1.A4.1 (p. 1 here) will make it redundant.

A5.1 *B* γ *r y* is *an att* ≡ *B r y* is *an att*

With '*B* γ *r y*' for '[ϕ(it)]', for a quantified object,

5.27 $\Theta\, x\, r\, \mathcal{E}\, b$ = $\Theta\, x\, r\, \mathcal{A}\, b$ & B a b is *predle* & $B\, r\, y$ is an *att* (5.5D)
that x is r of every b (that every b is $\gamma\, r\, y$)

This expands to the usual Fregean form, with conditions.

5.28 $\Theta\, x\, r\, \mathcal{E}\, b$ = $\Theta\, (u)(\sim u$ is a $b \vee x\, r\, u)$ & B a b is predle & $B\, r\, y$ is an *att* (5.8D)

5.29 $\Theta\, x\, r\, \mathcal{N}\, b$ = $\Theta\, (u)(\sim u$ is a $b \vee x\, n\, r\, u)$ & B a b is predle & $B\, r\, y$ is an *att* (5.6D)
that x is r of no b

5.30 $\Theta\, x\, r\, \mathcal{b}\, b$ = $\Theta\, (\#u)(u$ is a b & $x\, r\, u)$ & B a b is predle & $B\, r\, y$ is an *att* (5.7D)
that x is r of some b

These take quantified subject expressions in the usual way.

5.31 $\Theta\, \mathcal{E}\, c\, r\, \mathcal{E}\, b$ = $\Theta\, (v)(u)(\sim v$ is a $c \vee \sim u$ is a $b \vee v\, r\, u)$
& B a b is predle & B a c is an *att* & $B\, r\, y$ is an *att*
that every c is r of every b

5.32 $\Theta\, \mathcal{E}\, c\, r\, \mathcal{b}\, b$ = $\Theta\, (v)(\#u)(\sim v$ is a $c \vee (u$ is a b & $v\, r\, u)$
& B a b is predle & B a c is an *att* & $B\, r\, y$ is an *att*
that every c is r of some b

Others follow according to rule. 'The same' is unique to relations, with quantified object or separately mentioned objects. The underlining here restricts the notations to their proper contexts. Note how the order of sentential quantifiers differs from that of the infix quantifiers. This is another case where standard logics do not give workers *room to manoeuvre.*

5.9D $\Theta\, \mathcal{E}\, c\, r\, \underline{\mathcal{b}}\, b$ = $\Theta\, (\#u)(v)(u$ is a b & $(\sim v$ is a $c \vee v\, r\, u))$
& B a b is predle & B a c is an *att* & $B\, r\, y$ is an *att*
that every c is r of the same b

5.10D $\Theta\, x\, \underline{\&}\, y\, \underline{\&}\, \ldots\, r\, \underline{\mathcal{b}}\, b$ = $\Theta\, (\#u)(v)(u$ is a b & $(v\, r\, u \equiv v = x \vee v = y \ldots)$
that x and y and ... are r of the same b

Unlike 'every', 'any' is non-committal regarding import.

5.33 $\Theta\, \mathcal{A}\, b\, r\, x$ = Θ wti n a $b : r\, x$ & B a b is an *att* & $B\, r\, y$ is an *att* (5.4D)
that any b is r of x

5.34 $\Theta\, x\, r\, \mathcal{A}\, b$ = Θ wti E [it is n a $b : x\, r$ it] & $B\, r\, y$ is an *att* & B a b is an *att* (5.4D)
that x is r of any b

Next, the superfluous parentheses show the scope of the predicate negation.

5.35 $\Theta\, x\, n\, (r\, \mathcal{A}\, b)$ = $\Theta\, x\, Z$ & $\sim x\, r\, \mathcal{A}\, b$
that x is not r of any b
= $\Theta\, x\, Z$ & $\sim$ wti E [it is n a $b : x\, r$ it] $\vee \sim B\, r\, y$ is an *att* $\vee \sim B$ a b is an *att* (5.34)

This allows affirmation of the predication that Russell is not hezabodic of any treblott or any unicorn or any horse and is not a foal of any treblott or any unicorn or any horse. The negation is a protection against attempted affirmation of gibberish, authorial liberty determines sooth or nonsooth for non-exemplified attributes, and subject matter determines truth or falsity regarding zons. So I can tell my granddaughter that (*fictively*) the pony in my hall cupboard *is* a foal of some unicorn. But *it is not the case that* the pony in my hall cupboard is the foal of any unicorn (for lack of any unicorns). *It is not the case that* the pony in my hall cupboard is the foal of any horse (there is no such pony). The latest winner of the Derby *is not* a foal of any unicorn (for lack of any unicorns). The latest winner of the Derby *is* a foal of some horse.

The versatility (vagueness) of 5.35, so useful for literary purposes, makes assertions too unspecific for veridical intellectual purposes. Here I take a linguistic liberty to correct the lack. I call on 'nay' (<u>n</u>) in the role of 'not' restricted to the descriptive part of a verbal relational expression, not extending to any objects, so that expressions for objects are not in its scope, as they are for 'not'. 'Nay' is needed only for active verbs (is nay looking for). 'Non-' is available for nouns ('a non-participant in') and 'un' is available for participles ('unrelated to'). Though I offer colloquial expressions here, full development of these linguistic quirks is deferred, not being needed in the limited range of matters treated in the present stage of SRD. Again, parentheses are merely initial informal guides because the syncategorematic notations have built-in scopes.

5.11D Θ *x* <u>*n*</u> (a *c*) Ⱥ *b* = Θ *x n* (a *c* Ⱥ *b*) & *B* a *c y* is an *att* & *B* a *b* is an *att*
that *x* is a non-*c* of any *b*

5.12D Θ *x* <u>*n*</u> (*r*) Ⱥ *b* = Θ *x n* (*r* Ⱥ *b*) & *B r y* is an *att* & *B* a *b* is an *att*
that *x* is un-*r* of any *b*

This allows assertion of the predication that Russell is nay seeking, is a non-cousin of, and is unrelated to, any unicorn, which does not prejudge the availability of any unicorn, but requires seeking *y*, being a cousin of *y* and being related to *y* as attributes. By 5.5D, call on predicability provides treatment of 'every', not just 'any'.

5.36 Θ *x* <u>*n*</u> (*r*) ℰ *b* = Θ *x n* (*r* Ⱥ *b*) & *B r y* is an *att* & *B* a *b* is *predle*
that *x* is un-*r* of every *b*

This permits the assertion, as true, of the proposition that Russell is (genealogically) unrelated to every horse, which does prejudge the availability of some horse.

It can be seen that the vernacular, which SRD merely copies, with a few neologisms added for precision, is permeated with indications of zulogical import, providing a realm of logically important content, and that negations with various scopes are the key to distinctions in zulogy. Applied to the material world, the distinctions protect against purported realistic discussion of phantasms, and applied to attributes and predications, they protect against attempts at meaningless assertions. Thus standard logics are cut off from both developments, not in their elements, because Frege's basis is adequate and sound *as the elements*, but because the logics have not been elaborated in the right direction, through misguided blatant disregard of ordinary language.

All this treatment of quantification of the objects of relational predicates should not be confused with Hamilton's doctrine of the "quantification of the predicate", which is different again from quantification of predicates in higher-order conventional calculi. The apparent aim of the latter, incoherently treated conventionally, is achieved by ordinary quantification of attributes.

Classification of quantified predications

I shall return to further important applications of quantifiers later, but now it is time to catch up on metatalk. I formalize the basis of the judgment that universal and "existential" predications are logically useless, zulogical predications are useful, and general and particular predications are logically versatile and powerful.

5.13D *B* an *opd* = *B* = Θ wti *f*
being an open predication

5.37 Θ wti *f* is a *predcn* ⊃ Θ wti *f* is an *opd*

5.14D *B* a *zpd* = *B* = Θ tia *b*
being a zulogical predication

5.38 Θ tia *b* is a *predcn* ⊃ Θ tia *b* is a *zpd*

5.15D *B* a *gpd* = *B* = Θ ℰ *b* is *f*
being a general predication

5.39 Θ ℰ *b* is *f* is a *predcn* ⊃ Θ ℰ *b* is *f* is a *gpd*

5.16D *B* a *ppd* = *B* = Θ ƀ *b* is *f*
being a particular predication

5.40 Θ ƀ *b* is a *f* is a *predcn* ⊃ Θ ƀ *b* is *f* is a *ppd*

Conventional "existential" formulas, being *completely empty* due to negation, are not worth formal definition. As already shown, a relational predication may be multiply general or zulogical. It can even be mixed.

Traditionally a subject-quantified predication is taken as universal or particular but it is now seen to be general or particular. Modern logics fail to cope with syllogistic because what they call universal and existential are actually open and "existential". But SRD provides two classifications. One divides these predications into open and zulogical, the other into general and particular. Open and zulogical predications are non-polarized and, as the raven example shows, have a certain indefiniteness as to just what is being said. "The whole universe" (whatever universe there is, none being claimed) is relevant to them, as the associated expressions testify: whatever there is (never mind where or when) ...; it is not the case that whatever there is (never mind where or when) ... . They do not aid in the rational directing of discussion or in the cumulation of accepted results. A zulogical predication, unlike an "existential" one, succeeds in making an ontological claim. General and particular predications, being polarized, provide a definite indication of what is "old" or taken for granted, and of what is "new" or being put up for consideration. In each case a definite portion of the universe is relevant and the rest can be ignored. The means of direction and cumulation are provided. Theorems bearing on this informal discussion are provided below.

Universal and singular predications

A universal (without quibble) predication, calling on the whole universe, and claiming a universe to call on, is a general predication about *zons*.

5.17D *B* a *upd* = *B* = Θ ℰ *zon* is *f*
being a universal predication.

5.41 Θ ℰ *zon* is *Arist* is a *upd* (see 1.5.2D)
That every zon is Aristotelian is a universal predication

There is a traditional question whether a singular statement is "universal" or "particular", treating these terms as polar. In SRD there are two questions (traditionally conflated) with opposite answers. Is a singular predication universal or zulogical? Neither, since it is polarized and they are are not. Is it general or particular? Both, because it is "about" "all the items", identically "some of the items" exemplifying a definite description specifying a unique item (1.6.5D and 1.6.6D). According to some mediaeval opinion, Aristotle is the (pre-eminent) philosopher.

a	Aristotle is *t* philosopher	
b	Aristotle = τ philosopher	
c	τ philosopher = Aristotle	
d	Aristotle is mortal	(singular)
e	τ philosopher is mortal	(singular)
f	ℰ *zon* θ = τ philosopher is mortal	(general)
g	ƀ *zon* θ = τ philosopher is mortal	(particular)

5.18D *B* a *spd* = *B* = Θ τ *b* is *g*
being a singular predication

5.42 Θ τ *b* is *g* is a *spd* ≡ Θ τ *b* is *g* is an *predcn*

As just shown, a proper name may be called on for the definite reference.

MEANING RELATIONS

It is familiar that a proper treatment of language requires discourse on an utterance, a meaning, a user of the utterance and a recipient of it. Treatment of user and recipient calls on psychological matters. A great deal of work must be done in SRD before this general situation can be approached. As already forecast, utterances, mainly textual rather than vocal and without regard to meanings, will be treated in a later volume. Perhaps the formal foundations of psychology will also be treated, but the present problem demands a substantial development of psychology beyond its foundations. So here I give an informal account of these matters, one that I hope can be formally captured at a suitable stage. The only matters that can be formally treated now are those to do with meanings (attributes and predications) alone.

Ironically, the topics treated with rigour here are those treated with the most confusion in extant Logic and Linguistics, and the topics deferred here as too difficult are treated elsewhere with apparent confidence, but in vastly different ways by different authors. The examples in 4) p. 32 provide a sample of naive techniques.

Expression, reference, use, mention, affirmation ...

By use of a sentence it can be queried *whether Socrates is a logician*. If Socrates is a logician then certain conclusions may be forthcoming. In the context of 'whether' or 'if' the sentence expresses a predication about Socrates, but there is no affirmation or denial of that predication. "Is it six o'clock yet?" does not tell you the time.

1) A sentence *expresses* a predication. What predication? In the present case: Θ Socrates is a logician. In another case: Θ Pegasus is a flying horse. When I say that (Θ Socrates is a logician) is a predication or I say that (Θ Pegasus us a flying horse) is a predication, I am not talking about Socrates and I could not possibly be talking about Pegasus since there is no such thing. I am talking about a predication that successfully purports to be "about" Socrates or one that unsuccessfully purports to be "about" Pegasus. But I am neither affirming nor denying that proposition. (I have from the start in SRD said informally that a predicate *expresses* an attribute.)

2) I use 'mention' in the sense that in saying that the commander of the vanquished at Waterloo was sent to St. Helena I mention Napoleon, without *mentioning* (but *using*) his name, though in saying that 'Napoleon' has eight letters I mention Napoleon's name (by using a quotation of it) without mentioning Napoleon. (If only other people could heed Quine's advice on this matter, everything would then be clear and familiar, except for the Platonic error (see Volume 5)). A 'Θ' clause *refers to* a predication. I cannot *mention* any item without *using* some *utterance* that *refers* to it. In ordinary dialogue a sentence uttered separately indicates *affirmation* of its predication — in being affirmed, the predication is inevitably concurrently expressed.

Socrates is not a logician

(Prefixed with a '*B*', a predicate refers to the attribute expressed by the predicate.)

3) A sentence may be used to *affirm*, as well as to express, its predication, hence to deny the negation of that predication.

In the following, 'Q' is a dummy quantifier replaceable by '𝒜', 'ℰ', '𝒮' or '𝒩'. This is an informal expository procedure, with no systematic relevance.

1) A suitable case of '*Q b* is *g*' expresses a predication. What is the condition? According to 5.4D,

Θ𝒜 *b* is *f* = Θ wti *n* a *b* : *f* & *B* a *b* is an *att* & *B f* is an *att.*

Part of the condition is that each side of 5.4D refer to some (the same) predication. So it is required that Θ *B* a *b* is an *att* be a predication. It is not relevant whether or not *B* a *b is* an attribute. 5.4D makes no reference to the attribute *B* a *b*. From 2.1,

a *B* = *y* is an *att* & *B f* is an *att* ⊃ Θ *y* is *f* is a *predcn*

b *B* = *B* a *b* is an *att* & *B* an *att* is an *att* ⊃ Θ *B* a *b* is an *att* is a *predcn*

The latter conjunct is a theorem so the residual condition is the former. This is so when *B* a *b* = *B* a *b*, when *B* a *b Z*, hence when *B* a *b* is an *att.*

2) 'Θ *Q b* is *g*', in suitable cases, *refers to* a predication.

3) 'Every man is mortal' may be used to *affirm* Θ ℰ man is mortal ω a predication. By 5.11,

a Θ ℰ man is mortal is so ⊃ Θ *B* a man is *predle* is so

Note that, in contrast with 1 above, I now require *being so,* not just being a predication.

b ℰ man is mortal ⊃ *B* a man is *predle* (*a*, 2.13)

c ⊃ (#*u*) *u* is a man

d ⊃ 𝒮 man *Z*

Note that I still have not made any categorical assertion. Parallel with *b*, there is *e*. 'Every unicorn is a horse' may be *used* to *affirm* Θ ℰ unicorn is a horse ω a *predcn.*

e ℰ unicorn is a horse ⊃ B a unicorn is *predle*

f ⊃ (#*u*) *u* is a unicorn

g ⊃ 𝒮 unicorn *Z*

Point 4, in the next chapter, takes up the thread from here. In the meantime, SRD is developed to deal with old topics in new ways.

'That'

A lot hangs on the attitude taken to 'that' in the vernacular. How is 'Mary said that the television is broken' to be divided? In SRD, as just shown, 'that the television is broken' fills the blank in 'Mary said *x*', entirely in keeping with ordinary usage. Mary did not say nothing; she surely said *something*. Though many informal discussions call on similar cases, often with the insertion 'the proposition' or 'the statement' inserted before 'that', when it comes to formal work, the division is commonly made as follows. 'Mary said that' 'the television is broken. The 'that' is swallowed up in the 'said' phrase, so that a sentence replaces the "sentential variable" in 'Mary said that *p*'. By an elision that is correct but is becoming so common that it is no longer seen as an elision, 'that' is omitted, giving 'Mary said the television is broken' and, in general, '*x* says *p*'. Given that Mary said something, 'the televison is broken' is then wrongly taken as a referential expression. But the sentence does not *refer to* the predication, it *expresses* it. Grammatically, if '*p*' marks a sentential blank, then a filling is a clause; but a filling for 'that *p*' is not a clause, it is a noun phrase.

Without much thought's being expended, 'statement', 'assertion', 'proposition' and such are taken in one place as synonymous with 'predication', for what is said, and in other places as synonymous with 'sentence', for what is spoken. So arises the ubiquitous '*p* implies *q*', in PM, in Lewis's modal logic, in mathematics and in careless philosophy. Usually writers who work this way do not suppose that a sentence *does* refer to something, though the distinction between expressing and referring is not too widely recognized. But Frege, to the puzzlement of later commentators, specifically declared that a sentence *is* referential, that it refers either to *the True* or to *the False*. And in general he held that a proper name both expresses its sense and stands for or designates its referent. Given this, 'Russell' refers to the man Russell and (incoherently) expresses the attribute being identical with Russell, or some attribute taken as equivalent. If such an acute thinker as Frege could be so confused about this issue, it seems that the distinction is not so simple as I have suggested it is. But simple or difficult, it is a distinction that lies at the heart of language and logic. To flout it is catastrophic.

This distinction is parallel to Quine's use (expressing) *vs* mention (referring). Though he acknowledges referring (though taking a rosy view of it, as I show in Volume 5) he will not countenance expressing because he recognizes that there must be an object for the relation, which cannot be anything but a meaning (an attribute or a predication), neither of which he would give house room. So he calls on persons, users of languages, who *use* utterances and *mention* things, thus informally entering an arena whose rigorous discussion, as already said, is a long way off for SRD.

Base and predicatum

SRD is now so developed that it can be brought to bear on more questions that continue to trouble some people and that have been dismissed as pseudo-questions by others. As in other cases there are opinions that are easily reached informally, and that work, but these have not till now been substantiated by formal logic.

6.1D *B base y* = [$y = \Theta\ Q\ b$ is g & it $= B$ a b]
being a base of *y*

6.2D *B predtm y* = [$y = \Theta\ Q\ b$ is g & it $= B\ g$]
being a predicatum of *y*

I use the neologisms 'base' and 'predicatum' because 'subject' and 'predicate' are used in logic and grammar for notations that *refer to* a topic or *express* an attribute.

a *x base y* ≡ $y = \Theta\ Q\ b$ is g & $x = B$ a b

6.1 *B* a *b base* $\Theta\ Q\ b$ is g ≡ $\Theta\ Q\ b$ is $g\ Z$ & B a $b\ Z$

a *x predtm y* ≡ $y = \Theta\ Q\ b$ is g & $x = B\ g$

6.2 *B g* is *predtm* Θ *Q b* is *g* ≡ Θ *Q b* is *g Z* & *B g Z*

6.3 *B* a *b* is an *att* & *B g* is an *att* ⊃

B a *b base* Θ *Q b* is *g* & *B g* is *predtm* Θ *Q b* is *g*

Note that there is a predication for every sortal attribute with every attribute; it is not needed that any particular one (of infinitely many) be extant before being referred to.

In the definiens for 'Θ ϕ(ℰ *b*)' and for the other quantifiers that have zulogical import (5.5D – 5.7D), the predicability of *B* a *b* is explicitly claimed, by use of 'is'. Since Θ ϕ(ℰ *b*) is the same predication as that expressed by the definiens, it involves that claim. But the utterance containing the vernacular quantifier does not have a copula to indicate the claim. Because of ellipsis, the function of the copula has been bound up in the complex role of the quantifier. Here predicability of *B* a *b* is assumed or, to use the standard expression formally introduced below, presupposed. By contrast, the 'is' in '*Q b* is *g*' in these cases explicitly expresses predicability of the predicatum *B g*, in regard to such exemplars of the base as the quantifier selects: any one, every one, some one, a definite one or none of them. A singular predication Θ τ *b* is *g* or Θ *x* is *g* is covered because of its identity with a certain general predication.

Analytic and necessary predications

I am not so far, in SRD, concerned to deal explicitly in detail with necessity. Suffice it that all and only theorems would qualify as necessarily affirmable in an explicitly modal presentation. It is sometimes claimed or assumed that analyticity amounts to necessity but for SRD that "claim" is incoherent. Necessity has to do with affirmability but analyticity has to do with meaning. A number of issues have been confounded in disputes, this being inevitable when there is no calculus to keep discussion on solid ground. An analytic statement is said to be 1) true by virtue of meaning, hence true, or 2) true by virtue of meaning, if true at all. But there does seem to be an interesting notion, distinct from necessity, that is worth crystalizing. That is Kant's, sometimes expressed by saying that 3) the concept of the subject is included in the concept of the predicate. But what is a concept, what is the subject, what is the predicate, what is containment and are subject and predicate notations or referents?

SRD's answer combines 2) and 3), the latter by definition, the former by inference, though sooth takes the place of truth, thus permitting analytic fictional assertions and analytic falsities. The base is taken as the "concept of the subject", and the predicatum is taken as the "concept of the predicate" and containment in *y* is being a conjunct of *y*. Subject and predicate are nowadays pieces of text. The former is referential and the latter is expressive, showing that the traditional arguments are incoherent.

6.3D *B analt* = [(#*u*)(#*v*)(*u base* it & *v predtm* it & *v conjt u*)]
being analytic

a *B g conjt B a b* ⊃ *B a b* is an *att* & *B g* is an *att* & *B g conjt B a b*

b *B g conjt B a b* ⊃ Θ *Q b* is *g* is a *predcn* & *B g conjt B a b* (2.1D)

6.4 *B g conjt B a b* ⊃ Θ 𝒜 *b* is *g* is *analt*

6.5 *B g conjt B a b* ⊃ Θ ℰ *b* is *g* is *analt*

6.6 *B g conjt B a b* ⊃ Θ 𝔫 *b* is *g* is *analt*

6.7 *B g conjt B a b* ⊃ Θ 𝓈 *b* is *g* is *analt*

6.8 *B g conjt B f* ⊃ Θ τ *zon* θ *f* is *g* is *analt*

6.9 *B g conjt B f* & *B* = *x* is an *att* ⊃ Θ *x* ω *f* is *g* is *analt*

In all these cases, being so is irrelevant. It does not alter analyticity if there is no *b*, if there is no *b* that is uniquely *f*, if nothing is identical with *x*, or if *x* is not *f*. Analyticity as here defined is concerned purely with meanings, not with affirmations.

In the absence of lexical definitions, analytic predications are utterly banal.

a *Bf* is an *att* & *B g* is an *att* ⊃ *B g* is *conjt B f* . *g*

6.10 *Bf* is an *att* & *B g* is an *att* ⊃ Θ *Q zon* θ *f* . *g* is *f* is *analt*

6.11 *Bf* is an *att* ⊃ Θ *Q zon* θ *f* is *f* is *analt*

To see that analyticity is independent of truth, consider that

6.12 *Bf* is an *att* ⊃ Θ 𝔫 *zon* θ *f* is *f* is *analt*

That no *zon* that is *f* is *f* is analytically false, judgable false by virtue of meaning, no matter how the world is. It is false (by virtue of meaning) that no horse is a horse, though there are horses. It is not so (by virtue of meaning, as well as for lack of referent) that no unicorn is a unicorn, though it is not the case that there are any unicorns.

The main interest in analytic predications lies in those that stem from lexical definitions, as in the following stock example.

D *B* a bachelor = *B* male . adult . *n* married

B male *conjt B* a bachelor

Θ *Q* bachelor is male is *analt*

The notions of a "semantic marker" and of a "semantic component", so popular these days, are presumably related to being a conjunct — exactly how, it is not for me to say. I remark merely that 'male' and 'being male', and all typographical equivalents, are ("concrete") notations, that being male is an attribute, an intension, a "meaning", an "abstract" object, and that there is no room in SRD for any relevant non-"concrete", non-intensional object, whether a "component", a "structure", a "concept" or other metaphysical figment. Such imprecise talk flourishes only because of the absence of the disciplined talk that SRD employs. As SRD clearly shows, there is nothing wrong with metaphysics; it is the figments emerging from undisciplined discussion that are objectionable. These arouse Quine's warranted ire, but his loyalty to conventional logic promotes his disloyalty to common sense. That is the root of the difference between us, as I once told him after he corrected a technical error of mine.

Essence, property and accident

As in the case of analyticity (and indeed generally) I am not concerned to defend any dogma regarding essence. Again, certain notions stand out as interesting regardless of how they are discussed, but to discuss them I must make some choice of standard or novel expression. Not every expression has, like 'exist', been so illtreated as to be beyond repair.

I do not take individuals as having intrinsic properties — that variety of essentialism is incoherent. One does not talk about a property of Jack Smith or of Napoleon — rather, something is a property of a plumber, a general, a philospher, etc.

6.4D *B t ess* a *b* = *B* = *B* a *b*
being the essence of a *b*

6.13 *B* a *b Z* ⊃ *B* a *b* is *t ess* a *b*

B a horse . horned is the essence of a unicorn

B male . adult . human . not married is the essence of a bachelor.

'Of the essence' is elliptical for 'a conjunct of the essence', i.e. for 'a property of'.

6.5D *B* a *propty* a *b* = *B conjt B* a *b*

The conjunction shows the affinity of analyticity to properties.

Granted certain definitions, for the sake of a familiar illustration,

Being rational is a property of a mathematician.

Being two-legged is a property of a cyclist

nonetheless, although Professor B. H. Neumann, FRS, is a mathematician and a cyclist, no conclusion follows that either attribute of B. H. Neumann is a property of him.

That B. H. Neumann is a cyclist is contingently so, "accidental" in Aristotle's first usage in *Metaphyisics,* Δ XXX. That matter can be adequately treated without using the word 'accident'. But his second usage is relevant to the question what is, and what is not, a reliable inductive predication. Pat and Mick, and no one else, occupy a certain bench at a certain time — say each exemplifies *B* sit. But Kruschev (this dates the example) might have sat on that bench at that time. Though everything that sat on the bench was Irish, yet had Krushchev so sat he still would not have been Irish. *B* Irish is an accident of a "sitter". Note that being Irish is generally predicable of "sitters", but is not a property of a "sitter". It is entailed by *B* sit, but neither necessarily nor "by virtue of meaning".

6.6D *B accdt* a *b* = [*B* a *b* ⇛ it & it *n propty* a *b*]
being an accident of a *b*

6.14 *x* is *accdt* a *b* ≡ *B* a *b* ⇛ *x* & *x n propty* a *b*

Since having its angles total two right angles is predicable (by theorem hence by necessity) of a (Euclidean) triangle, having its angles total two right angles is *necessarily an accident* of a triangle — that is Aristotle's very example, which some may have, up to now, considered an aberration. *B* Irish is contingently an accident of a "sitter". The problem of formulating a scientific law involves finding yet another kind of generality.

PROPOSITIONS

According to 5.5D and 5.6D, that being a unicorn is predicable is a conjunct in the definiens of that no unicorn is savage. Whether or not being a unicorn is predicable is not at issue. The same applies to being a horse and that no horse is savage. Their being predications has no regard to what is so or what is not so. Both predications are intelligible (being intelligible *is* just being an intension); one can as easily spot a unicorn as spot a horse, should either come one's way.

But one (an important one but not the only one) of the uses of language is as a means of discussing the world (and not merely the geographical world). If we consider the world we find overwhelming evidence that there are horses, that some horse zu, that being a horse is predicable, and no convincing evidence that being a unicorn is predicable. This non-linguistic difference reflects on the predications that no horse is savage and that no unicorn is savage. The former is "about" subject matter that constrains rational judgment; the other has not been shown to be "about" anything. In a word, the former is a *proposition* and the latter is not. How is this distinction to be formalized?

... and confirmation

As stated in the previous chapter,

1) A sentence *expresses* a predication;

2) A 'that' noun phrase *refers to* a predication;

3) A sentence may be used to *affirm* a predication.

So Mummy says to her toddler daughter "Santa Claus brought you a present.". *There* is an affirmation, of a predication that even a child can grasp. Clearly meaningfulnes and affirmation are independent of actual relevant subject matter. But affirming a predication does not make it so, except in fiction. Point four is that

4) Subject matter may *confirm* (or disconfirm) a predication.

We have seen that a predication involves a base and a predicatum. To test whether no horse is savage we must consult the world, examine those things (horses) that exemplify the base (being a horse) and see, for each one, whether it exemplifies the predicatum (being savage). The world decides for us. Likewise, to test whether no unicorn is savage we must consult the world; but when we do, we cannot find any things (unicorns) that exemplify the base (being a unicorn), so the further step of testing for being savage never arises. The world decides this for us, too.

What, then, make the world relevant to one predication and not to another? The predicability of the base. What makes a world-relevant predication (a proposition) so or not so (true or false)? The "applicability" of the predicatum. What is taken for granted before we start the enterprise of consulting the world? That we are presented with a predication, something meaningful, something mentally graspable. Without that we do not know what to look for in the world. Once again it can be seen that "meaning comes before truth". I keep on repeating this obvious doctrine because so much of what logicians do totally disregards it.

Propositions

By this point it should be clear what a proposition is: being a proposition is having a predicable base. The following definition is provisional.

D *B* a *propn* = [τ *base* it is *predle*]
being a proposition

The horse/unicorn example is formalized as follows.

B a *h* is an *att* & *B sav* is an *att* & *B horned* is an *att*

D *B* a *uc* = *B* a *h* . *horned*

B a *uc* is an *att*

Θ ɳ *h* is *sav* is a *predcn*

Θ ɳ *uc* is *sav* is a *predcn*

B a *h base* Θ ɳ *h* is *sav*

B a *uc base* Θ ɳ *uc* is *sav*

B a *h* is *predle*

τ *base* Θ ɳ *h* is *sav* is *predle*

Θ ɳ *h* is *sav* is a *propn*

B a *uc* is *n predle*

τ *base* Θ ɳ *uc* is *sav* is *n predle*

Θ ɳ *uc* is *sav* is *n* a *propn*

In general:

7.1 *x* is a *propn* ≡ τ *base x* is *predle*

a *B* a *b* is *predl* & *B g* is an *att* ⊃ *B* a *b base* Θ *Q b* is *g*

7.2 *B* a *b* is *predl* & *B g* is an *att* ⊃ Θ *Q b* is *g* is a *propn*

7.3 *B* a *b* is an *att* & *B g* is an *att* & *B* a *b* is *n predl* ⊃
Θ *Q b* is *g* is a *predcn* . *n* a *propn*

Having a base at all makes a proposition a predication. Boolean compounds of propositions are thus predications, but they are not yet certified as propositions. So the full definition, replacing **D** above, calls on Booleans.

7.1D *B* a *propn* =
[τ *base* it is *predle* ∨ *x* is *agg propns* & it *P-neg x* : *P-conj x* : *P-alt x*]

"Subject"

Specifying propositions in terms of a base attribute rather than any exemplar of it provides a uniform treatment for all propositions. It avoids awkward questions regarding what a proposition is "about" and what is the "subject" of a "statement" in traditional syllogistic. Is that some horse is a unicorn really about (without scare quotes) horses — which ones? It is surely not about all horses, and it cannot be about those that are unicorns, since there are none. That no horse is a unicorn cannot be about no horse. What a predication is "about", involving subject or topic, except in the case of mentioned individuals, is an invitation to confusion that SRD declines.

a *x* Z & *B g* is an *att* ⊃ *x* = *x* & *B g* is an *att*

b ⊃ *B* = *x* is *predle* & *B g* is an *att*

c ⊃ Θ ɓ *zon* θ = *x* is *g* is a *propn*

d ⊃ *B* = *x base* Θ ɓ *zon* θ = *x* is *g*

e ⊃ Θ *x* is *g* is a *propn*

7.4 *x* Z & *B g* is an *att* ⊃ *B* = *x base* Θ *x* is *g*

Likewise

7.5 *x* Z & *B t b* is an *att* ⊃ Θ *x* ω *t b* is *g* is a *propn* & *B t b base* Θ *x* ω *t b* is *g*

Presuppositions

That its base is predicable is a condition that a predication be a proposition. Such a condition is a presupposition, not affirmed but taken for granted.

7.2D *B presupn y* = [*y* is a *predcn* & it = Θ τ *base y* is *predle*]
being a presupposition of *y*

7.6 Θ *B* a *b* is *predle presupn* Θ *Q b* is *g* ≡ Θ *Q b* is *g* is a *predcn*

Failure of its presupposition to be so prevents a predication from being a proposition but does not alter its status as a predication.

That there is a lion follows from a presupposition of the predication that no lion is in the Bronx Zoo. Since there is a lion, the presupposition is true and the main predication is a proposition. Since there is (I suppose) a lion in the Bronx Zoo, that proposition is not so, hence false (see below).

That there is a unicorn is a presupposition of the predication that every unicorn is in the Bronx Zoo. That presupposition is false, so the main predication is not a proposition. Its affirmational status is discussed in the next section.

Responses to affirmation

There are four mutually exclusive responses that one can make to an affirmation in SRD. Expression of two of them demands neologisms, for the lack of which ordinary English is defective, to every honest person's disadvantage.

1) "Russell is a philosopher."

In a *concessive* response, acknowledging the presupposition that Russell zu, and conceding applicability of the predicatum: being a philosopher, one can say "Yes".

2) "Russell is a disciple of Kant."

In a *conservative* response, acknowledging the presupposition that Russell zu, but not conceding the applicability of the predicatum: being a disciple of Kant, one can say "No". The availability of "Yes" and "No", and their standard use, testify to the presumption that, unless circumstances indicate otherwise, people, whether right or wrong, are talking seriously, i.e., so as to express propositions.

3) "Russell, who is French, wrote *Mysticism and Logic*."

In a *rejective* response, rejecting the presupposition that Russell is French, and therefore disregarding the predicatum, but acknowledging that something intelligible has been affirmed, one can say "Ono".

4) "Every sentence has a deep structure."
"The concept blue is a value of the dimension colour."
"The concept *blue* is a value of the dimension *colour*."
"The concept "blue" is a value of the dimension "colour"."
"The concept 'blue' is a value of the dimension 'colour'."
"A relation is a set of ordered pairs."
"The disk holds 32 Megabytes of information."

In a *nonplus* response, where the utterance does not express any recursively grounded predication, but leaves one baffled, not knowing which of a variety of predications is intended, or wondering whether anything intelligible has been affirmed, one can say "Nono". My nonplus response to all pretentious efforts at theory and at theorizing about theory in the 20th century was the initial motivation for developing SRD, and is the reason for the proliferation of raised-eyebrow quotes in my work.

Definitions follow. A concession of a proposition is the proposition itself. This is independent of the various ways in which the proposition might be expressed.

7.3D *B concen y* = *B* = *y* ω a *propn*
being a concession of *y*

7.7 *y* is a *propn* ⊃ *y concen y*

Response 2 needs a different types of negation. Here '*Q*' can be either 'every' or 'some'. The primed value of 'every' is 'some' and that of 'some' is 'every'. Treatment of 'no' follows from its definition.

7.4D *B conser-neg* Θ *Q b* is *g* = *B* = Θ *Q′ b* is *n g* ω a *propn*
being a conservative negation of Θ *Q b* is *g*

7.8 Θ *Q b* is *g* is a *propn* ⊃ Θ *Q′ b* is *n g conser-neg* Θ *Q b* is *g*

Response 3), with '*Q*' unrestricted, denies that any proposition has been expressed.

7.5D *B rej-neg* Θ *Q b* is *g* = *B* = Θ Θ *Q b* is *g* is *n* a *propn*
being a rejective negation of Θ *Q b* is *g*

7.9 Θ *Q b* is *g* is a *predcn* & *B* a *b* is *n predle* ⊃
Θ Θ *Q b* is *g* is *n* a *propn rej-neg* Θ *Q b* is *g*

Note that 'Θ' is doubled, because the main predication has a predication for its topic.
Response 4) involves denial that any predication has been expressed.

7.6D *B nonplus y* = *B* = Θ ~ *y* is a *predcn*
being a nonplus of *y*

7.10 ~ *y* is a *predcn* ⊃ Θ ~ *y* is a *predcn nonplus y*

Thus the various responses, in explicit form, for a purported affirmation Θ *Q b* is *g*, are the following.

Concessive:	Yes. *Q b* is *g*
Conservative:	No. *Q′ b* is *n g*
Rejective:	Ono. Θ *Q b* is *g* is *n* a *propn*
Nonplus:	Nono. ~ Θ *Q b* is *g* is a *predcn*

Note how, in the first two cases, the response is expressed by an utterance lacking 'that', because the response is a simple judgment about subject matter. But in the latter two, the utterance includes 'that' because these judgments are meta-affirmations, in the one case denying that a predication is a proposition and, in the other, denying that any predication has been affirmed. No doubt the complication of the last two cases contributes to their absence from everyday talk. The proverbial "Are you still in the habit of beating your wife?" most commonly calls for the response "Ono".

Truth and falsity

I now formalize some points that have been taken for granted in the previous sections. According to Aristotle (*Metaphysics* Γ VII) *Falsehood consists in saying of that which is that is not, or of that which is not that it is. Truth consists in saying of that which is that it is, or of that which is not that it is not.* In the present context this account can only provoke a nonplus response. Does 'consists in' amount to '=' or what? Does 'is' correspond to 'zu' or is it a copula? If the latter, what is the suppressed predicate? What sort of item corresponds to '*that* which is'? Is *saying* relevant? Such questions would usually be carping criticism, when the critic had nothing logically better to offer. But in SRD, the same ground is covered with full rigour. All that remains is to formally define the words 'true' and 'false'.

It would be profitless to add to the torrent of undisciplined, or else Pickwickian, discussion of truth. In particular, it is a farfetched claim of Tarski's that Aristotle's account supports his "semantic" theory or truth or falsity of *sentences*. Sentences may be spoken but they cannot be *said*; Aristotle is concerned with something said, i.e. a predication, not with something spoken. At this stage of development of SRD, the nature of truth is pretty obvious.

7.7D *B* true = *B* a *propn* . so

7.11 *x* is a *propn* ⊃ (*x* is true ≡ *x* is so)

7.12 Θ*p* is a *propn* ⊃ (Θ*p* is true ≡ *p*)

7.8D *B* false = *B* a *propn* . *n* so

7.13 *x* is a *propn* ⊃ (*x* is false ≡ *x* is *n* so

7.14 Θ*p* is a *propn* ⊃ (Θ*p* is false ≡ ~*p*)

Now all the results for 'so' and '*n* so' for predications in general are carried over to 'true' and 'false', allowing the sentential negation to be replaced by predicative one.

SRD is not (yet) concerned with a full treatment of psychological *saying* by a person. Its concern is with logical *expression* by a utterance. No end of predications are sayable, waiting for someone to say them.

The SRD treatment makes propositions the "bearers of truth and falsity", but not "bearers by virtue of meaning". Being true or else false is an accident, not a property, of propositions. Because of this, SRD can deal with predicational content divorced from all modes of affirmation, so that meanings as such are discussable. Lacking this facility, 20th-century logic resorts to discussion of notations in order to escape the affirmational context. Without discussion of meanings, semantics is impossible and without a way of rigorously discussing meanings, modern "semantics", whether of the linguistic or the logical kind, is floundering.

The Liar

"Epimenides, a Cretan, said 'All Cretans are liars.'." "I am now lying." "What I am now saying is false." There are a variety of specimens like these, constituting the paradox of The Liar, which pose a problem for any attempt to define truth by undisciplined use of ordinary language. But paradox here rests on a vast, vague, unformalized logical and psychological theory of propositional (more generally predicational) attitudes, which SRD can only ignore.

Modern variants of the paradox are such as "This sentence is false." "The next sentence is true. The previous sentence is false." Certain versions of these can formulated with great precision and can be shown to to produce an antinomy, in company with a direct sentential analogue of 2.12D. Otherwise lacking an account of truth, and lacking any rigorous treatment of intensions that could lead to one, 20th-century logic rejects the direct analogue and resorts to roundabout methods of great complexity, *à la* Tarski, Quine, Carnap and others. But SRD, having a suitable groundwork laid, succeeds with 2.12D and 7.7D. According to these, the troublesome quotations above are all rejectable out of hand as incoherent, for in SRD (as, of course, in ordinary life), no sentence is either true or false.

Some plugs and a forthcoming attraction

Tarski's account suffers from the further fault of calling on set-theoretical methods, whose deficiencies have already been pointed out in Volume 2. Russell's two versions of his antinomy were examined and found faulty in *Definition, Russell's Antinomy, Etc.*, as a necessary preliminary to the whole enterprise of SRD. The third member of the Unholy Trinity is Gödel's theorem(s), with offshoots in Turing's work.

A single flaw is, of course, enough to discredit any purported rigorous demonstration, but Cantor's diagonal argument, another ingredient of Gödel's work, has been discredited inVolume 2. Other faults will be shown in Volume 6, when the full panoply of SRD is available. It indicates something rotten in the state of 20th-century Academe that so much faulty superstructure has been raised, *with no formal error* and with such waste of effort, on such a flimsy foundation, in a global approach that rests on so many intellectual aberrations. As *The Emperor's New Logic* explains at length, the causes are basic flaws of character: *a disregard of common sense* and *frivolous resort to game-playing in the face of difficulties*.

SOME APPLICATIONS

To further warrant the just made criticisms, the examples that follow support the claim that SRD, as it develops, is acquiring more and more of the terseness, flexibility and versatility of ordinary English, without departing from the strictest canons of formal logic. It is of technically logical interest that SRD is able to do this, but the wider philosophical and practical benefit is that the result overwhelmingly supports common sense and ordinary good judgment when that conflicts with the doctrines of logicians. Logicians are forced into their own problems because their techniques are not sufficiently developed to confirm obvious results.

Identity and necessity

I first of all get necessity out of the way, as irrelevant to questions of lexical meaning. In standard modal logic,

a $L\,x = y \supset x = y$

b $x = y \supset (\phi(x) \supset \phi(y))$

c $x = y \supset (L\,x = x \supset L\,x = y)$

d $L\,x = x \supset (x = y \supset L\,x = y)$

e $x = y \supset L\,x = y$

f $x = y \equiv L\,x = y$ (*a*, *e*)

There, every identity is taken as necessary. SRD, when modalized, differs in two ways. First, replacement for non-intensional identity in intensional contexts is prohibited, so *c* above is not warranted. Second, it is not a theorem that $x = x$, hence not that $L\,x = x$, so *e* fails. SRD thus permits affirmation of non-necessary identities.

Accidental identities

The Ancient Greeks gave the name 'Hesperus' to a certain celestial body referred to, for particular reasons, as the evening star (τ *evestar*). In accordance with that usage it is nowadays agreed that

Hesperus = τ *evestar*, i.e. Hesperus is *t evestar* (see Volume 1)

Independently they gave the name 'Phosphorus' to a certain celestial body, referred to, for certain (obviously different) reasons, as the morning star (τ *mornstar*). Accordingly

Phosphorus = τ *mornstar*

In view of the different circumstances, it would be silly to suppose that the attribute predicable of Hesperus, *B t evestar*, is the very same attribute, *B t mornstar*, predicable of Phosphorus. The attributes are independent of examplars, and each might have been defined merely on the prospect, not confirmed, of finding a single initial sign of the end of the day and a different single initial sign of the end of the night.

B t evestar ≠ *B t mornstar*

Even if the Greeks had believed that the same celestial body was involved on both occasions, they would still not have presumed a single attribute. Further, if they had so believed, they would have had no reason to giving a single thing two names, even if it fitted two descriptions. Their presumption. like everyone else's for 2,000 years, was that Hesperus ≠ Phosphorus.

But progress in astronomy has shown them to be mistaken. As we now know,

Hesperus = Phosphorus = Venus.

It follows that

Hesperus = τ *evestar* = τ *mornstar*

Phosphorus = τ *evestar* = τ *mornstar*

τ *evestar* = τ *mornstar*

A single body, Venus, is both the morning star and the evening star. There is one thing with two attributes, and now three referential notations.

These conclusions just happen to be true; they are not necessary (not theorems) and not analytic (dependent on meaning). The Greeks might have assigned other names in the first place but, given that they used the names indicated, and that English is what it is, and SRD is what it is, the conclusions are true just because the world is as it is.

Because a nominal predication is a general predication, and here the base entails but is not a conjunct of the predicatum (6.6D),

B = Phosphorus *accdt* of a *zon* that = Hesperus,

So, that Hesperus = Phosphorus is a case of accidental identity. It is interesting that in such cases, Aristotle's two interpretations for 'accident' (as not necessary and as not a property) are equivalent (*Metaphysics* Δ XXX).

Essential identities

It is known that

Cicero denounced Catiline.

Further,

Tully = Cicero, so Tully denounced Catiline.

But whereas astronomical research was needed to show that Hesperus is Phosphorus, no non-linguistic historical research about Cicero's life and times is needed to prove that Tully denounced Catiline, given that Cicero did. Whatever Cicero did, Tully did; whatever Cicero was, Tully was. Unlike 'Hesperus' and 'Phosphorus', 'Cicero' and 'Tully' are not two names originally given to what was only subsequently seen to be the same object. There was never any suggestion of two persons when an Ancient-Roman infant was dubbed Marcus Tullius Cicero. How is this to be expressed in SRD, without recourse to the mention of names?

Hesperus = Phosphorus

Tully = Cicero

No difference shows itself here. The difference is not that contingently Hesperus = Phosphorus but necessarily Tully = Cicero. It is false that necessarily Tully = Cicero, that necessarliy Cicero = Cicero, that necessarily Cicero zu. All these are contingent. As so often here, necessity is not to be confounded with meaning.

'Cicero', as a proper name, has no meaning, does not express an intension — it has a referent, Tully. Whereas Hesperus is *known to be* Phosphorus, it is not *known as* Phosphorus, but Cicero *is known as* Tully. By an implicit definition made at Cicero's naming,

D B = Tully = B = Cicero.

This definition of an identity attribute provides that Cicero = Tully (given that Cicero Z) but the extensional identity for Hesperus and Phosphorus does not provide a corresponding intensional identity. By D,

B = Tully is *t ess* σ *zon* θ = Cicero.

This is a case of essential identity. Note that it is not said that something is the essence of *Cicero*. Cicero $\rightarrow$ σ *zon* θ = Cicero (is quasi-identical with), but Cicero $n =$ σ *zon* θ = Cicero. For details, see Volume 1.

Not to know that the evening star is the morning star betrays ignorance of astronomy. Not to know that Hesperus is Phosphorus betrays either that or ignorance of certain naming conventions. Not to know that Cicero is Tully also betrays ignorance of certain naming conventions. But the latter two cases are distinguished by the differing origins of the various conventions.

Necessary identities

As already explained in Volume 1, a definition has no zulogical import.

$x = y \supset D\, x = y \;\&\; x\, Z$

In an explicitly modal calculus,

$L\, x = y \supset L\, D\, x = y \;\&\; L\, x\, Z$

In the case of an explicit definition, having axiomatic force, the first conjunct is valid, so

$L\,(x = y) \supset L\, x\, Z$

Thus there are necessary identities only for "logical" *zons* like B an *att*, [it Z], Θ B *att* is *predle*, introduced by an axiom or by a definition having the force of a theorem. The only non-intension for which there can be necessary attributes is J. L (J = J) (see Volume 1) Otherwise, no attribute or exemplar gives a case of L (x = x). Note that D above is not dignified with a theorem number. It is not a theorem. But B an attribute is defined in a formally numbered definition, and necessarily it zu.

Hence, although D(B = Cicero = B = Tully), it does not follow that L D B = Cicero = B = Tully or D L Cicero = Tully or L Cicero Z or L Tully Z. Nor does it *yet* follow that B = Cicero = B = Tully or Cicero = Tully or Cicero = Cicero or Cicero Z, for lack of demonstrated referent. But given that Cicero denounced Catiline, or any other fact about Cicero, Cicero Z — then the *non-necessary* conclusions follow. It follows that Cicero = Tully, without recourse to further observation, whereas further observation was needed to show that Hesperus = Phosphorus, in the absence of a definition.

The zulogical point made here is illustrated by an identity for mythological figures.

D B = Zeus = B = Jupiter

No one was foolish enough to give the same use to two names. The Greeks assigned one, the Romans assigned the other, and the English borrowed both. But it is not the case that Zeus = Jupiter, though fictively Zeus = Jupiter.

Laws, induction and counterfactuals

It is shown on p. 17 that this

A ℰ raven is black

unlike this

B wti not a raven or black

satisfactorily expresses an inductive generalization regarding ravens, making ravens as such but not non-black things as such relevant. Also, given that only Pat and Mick exemplify *B* sit (i.e. occupy a certain bench at a certain time),

C ℰ *zon* θ sit is Irish.

Accordingly, being black is an accident (not a property) of a raven and being Irish is an accident of a sitter. So much for the *content* of these general propositions.

Now I consider warrant and consequences. These "indicative, material conditionals", which are not conditionals at all but alternations, are sound, though vacuous.

~ Polly is a raven v Polly is black

~ Kruschev is a sitter v Kruschev is Irish

Since Polly and Kruschev do not satisfy the required conditions, the consequents are not forthcoming. But it is intended that *A* should support a conterfactual conditional

If Polly were a raven then Polly would be black.

but that *C* should not.

* If Kruschev were a sitter then Kruschev would be Irish.

By intent, blackness "goes with" ravens in way that Irishness does not "go with" *zons* that sit. The difference may be informally indicated by a change of copula. Though we do not wish to "strengthen" *C*, we do wish to strengthen *A* to this,

D Every raven must be black.

which supports a counterfactual.

I am not concerned here with biographical and psychological matters (such as figure in the confused theory of statistical hypothesis testing) which might lead one to affirm *D*, nor with the strength of one's rational grounds for affirmations. My topic is neither scientific discovery nor the justification of induction, though these are important topics for discussion elsewhere. My problem, thoughout pure SRD, is simply *how to say it*. The first problem, and anybody's who seeks a rational account, must be to find a way to *express* an inductive generalization. Without a suitable formulation, the further problems cannot even be stated.

I next consider affirmability. *C* is exhausted by its instances. Pat is a sitter and Mick is a sitter, and nothing else is. *C* is a consequence of its instances. Pat is a sitter and Irish, Mick is a sitter and Irish, and nothing else is — that is why it is affirmable. It has no lawlike character to support a counterfactual, and one does not wish it to have any. But what about *A*? One is not saying merely that all examined ravens are black. That is true (or taken as true for the sake of the conventional example), is a consequence of its instances, and is not lawlike. But *A* itself jumps from examined ravens to ravens in general. Some of its instances concern as yet unexamined, unhatched and unconceived ravens, so it is not founded on them. Thus so far *A* is not affirmable at all; indeed no amount of confirmatory evidence about ravens can ever make it affirmable — that's the nature of open induction. So although *A* has the lawlike quality, one is not empowered to invoke it.

A would arise from this,

L Every raven is black.

taken as a postulate, in a frank leap beyond attested cases. As explained above, the *warrant* for such a leap is not now under examination. But the following is too strong.

L(wti *n* a *b* : *E* [ϕ(it)] & *B* a *b* is *predle* & [ϕ(it)] is an *att*)

L(*B* a raven is *predle*)

L(*B* black is an *att*)

It is not part of my intent to claim that there *necessarily* is a raven or an attribute being black. Necessity is needed, but not globally in the assertion. Nothing of the *content* of the assertion needs to be modified. A new *mode* of assertion, expressed by *a new copula*, has to be defined.

What is needed is a predication that signifies the inductive leap, has a lawlike quality supporting a counterfactual, but is not zulogically creative — as stressed on various occasions, zulogical axioms, like "existential" axioms, though formulable, are anathema in SRD. So, finally,

8.1D Θ Ⱥ *b* must be *g* = Θ *L*(wti *n* a *b* : *g*) & *B* a *b* is an *att* & *B g* is an *att*

8.1 Ⱥ *b* must be *g* ≡ Ⱥ *zon* θ *n g* must be *n* a *b*

8.2 Ⱥ *b* must be *g* ⊃ Ⱥ *b* is *g* (*Lp* ⊃ *p*, 5.4D, etc)

Here, despite the unpolarized character of the postulate, and the validity of the inference, Hempel's paradox of confirmation is avoided, since the matter of piecemeal practical confirmation has been *over-ridden*. But disconfirmation or invalidation is provided by a single case of a raven that is not black. Only the 'whatever' clause, the openly quantified item (with no zulogical commitment), is postulated. 8.1D cannot be phrased in PM, although the postulate can be expressed in isolation,

|- (*u*)(~ *u* is a *b* ∨ *u* is *g*)

where '|-' has the identical force of '*L*', to signify a (primitive or derived) thesis. But in PM and its derivatives, '|-' is (unfruitfully) viewed as metalinguistc and outside the calculus, so it can never occur inside a formula. Quine makes '|-' explicitly metalinguistic, an extra-systematic notation to accompany references to closed formulas.

In a suitable calculus (the foundation of SRD, treated in Volume 6, where necessity is analysed and developed) 8.1D has all the requirements. A sceptic regarding black holes can accept that if there were such things they would have such and such attributes, e.g.: *Any* black hole must be *f*. But he should not therefore strengthen this to say that *every* black hole must be *f* until he is satisfied that some black hole has been observed. The only way to determine putative attributes of hypothetical objects is by inference from postulates (hence the need for necessity) since it is not the case that any suitable objects present themselves for inspection. When one does, its character can be considered in relation to the imputed attributes.

The previous case, an *open* counterfactual, dealt with the total absence of any relevant zons. But an *individual* counterfactual is also available. Betelgeuse is not a black hole, but if it were, it would be *f*.

8.1D provides the meaning for an induction, but there is no precedent for inference with 'must'. There I provide an axiom for giving inferential force to an inductive postulate.

A8.1 Ⱥ *b* must be *g* & *x* is *n* a *b* ⊃ if *x* were a *b* then *x* would be *g*

I now give a rigorous treatment of the stock example. By inductive postulate,

a *L*(wti *n* a raven : black).

By predicability through observation

b *B* a raven is an *att.*

c *B* black is an *att.*

d Ⱥ raven must be black. (*a*, *b*, *c*, 8.1D)

We know there are ravens, so this case is different from the previous one. Here again is a case of an *individual* counterfactual.

e Polly is not a raven, but

f If Polly were a raven then Polly would be black. (*d*, *e*, A8.1)

Here inference can go further than it did for hypothetical black holes.

g *B* a raven is *predle.*

h ℰ raven must be black. (*d*, 5.5D)

This is the strongest law of all, having both counterfactual force by postulate and zulogical import by observation.

Accidents strengthened to properties

One may be less reluctant to acknowledge a convincing example of a green raven or a pink elephant than to accept as iron something that is not magnetic. Historically, iron was used for engineering purposes before its magnetic character was discovered. The discovery was an inductive inference, like the one about ravens. The ravens example is modern, replacing a similar case of white swans, because the discovery of black swans rendered the traditional one false. It would be a hardy or foolhardy man who bet his life on the blackness of all ravens. But I do not doubt that there are some (besides myself) who would bet their lives on the magnetic *property*, being magnetizable, not necessarily magnetized, (not just the *attribute*) of iron objects. If a thing is not magnetic then it is not iron, no matter how well it otherwise simulates iron. As shown below the bet is perfectly safe.

The initial inductive case is like that of the ravens. Suppose the attributes being iron and being magnetic are available. These might, these days, be defined on some theoretical basis in physics, though originally on some crude practical test and taken as primitive. SRD does not lay down the law regarding what might be taken as primitive. If natural kinds should happen to be among the primitives then their lexical notations will have a degree of vagueness determined by people's variations in judgments, which may in practice be quite small. Such vagueness poses no valid deterrent to the quest for formal discourse concerning meanings, and does not demonstrate the "impossibility of semantics". On the realization, based on theoretical inference, that magnetism is inherent in iron in a way that blackness is not inherent in ravens, one has to make a new definition. Being iron, up to now defined, or taken as given, without regard to magnetism one way or the other, is redefined to as to include magnetism. If a mistake should be made, one just re-redefines. That is why the bet is safe.

D *B* iron = *B* iron . magnetic

As a conjunct of the definiens (i.e. as a conjunct of *B* iron), magnetism is now a property, is of the essence, of iron.

Dispositions

Being magnetic and being flammable are traditional examples of dispositions. The problem is usually stated in the form: if this piece of iron had been placed near a magnet, it would have been attracted, or if this match had been struck it would have ignited. The force of a disposition comes from a counterfactual, and this can come from either an induction (as in the raven case) or a definition (as in the iron case). The former is characteristic of observationally based inductions, in natural history; the later is at home in theoretical natural science.

Fiction

However it may be in contemporary efforts at semantics, in SRD a predication can be grasped in the absence of confirmation or disconfirmation. Indeed, if it is not grasped there is nothing to be confirmed or disconfirmed. An utterance can be rejected as meaningless by "nono". A propositional affirmation can be accepted by "yes" or denied by "no". A non-propositional "affirmation" must be denied by "ono". So, in strict factuality, in plain discourse, without ifs or buts or other qualifications,

"Oliver Twist was an orphan."

"Ono, it is not the case that Oliver Twist was an orphan."

One does not say "Oliver Twist was not an orphan.". In this plain mode of discourse it is never affirmable that Oliver Twist is not *f*, for it is not the case that Oliver Twist *Z*.

In typical formal languages (those that really are languages and not "languages", if there are any), lacking zulogy, Θ Oliver Twist is not an orphan is not distinguishable from Θ it is not the case that Oliver Twist is an orphan. These systems are stuck in the true/false mode. Since their Pickwickian "meanings" arrive via "truth" and "falsity" of sentences there is no room for intelligible discourse for which truth and falsity are irrelevant. But the 'ono' sphere of SRD provides room for just such discourse, wherein lies fiction. (As Volume 6 will show, the above judgment framed in conventional terms is not correct. Such calculi are indeed two-valued, but the values are not truth and falsity.)

For infants, the onset of fiction is signalled by "Once upon a time" or "Let's pretend", and its end by "They lived happily ever after." or "That's enough of that!". Adults often rely on more subtle, contextual cues. The psychology of this is not my present concern, but it would not be possible at all if there were no 'ono' sphere. It is silly to deny fictional affirmations; they are not put forward as being in agreement with any external subject matter. It is equally silly to deny axioms laid down by a theoretician. In either case, to offer a denial is to misunderstand the author's intention; no denial will be brooked. What such an author says is so simply by virtue of his saying it, but it is neither true nor false and no *evidence* is needed one way or the other.

I have nothing to say regarding the logical or factual regulation of fiction (or uninterpreted axiomatic systems); anything that anyone said would be completely disregarded by their artistic creators, and rightly so. But literary history and the textual criticism of axiomatic systems are not, as such, literature or axiomatics. They do have some subject matter, viz., what the author said and the fact that he said it. Whether someone said something or not, regardless of what was said, it a matter to be judged on evidence, typically relying on documents.

What is needed now is to acknowledge explicitly fictional discourse by a definition calling on the notion of saying something, and to distinguish the tasks of literature and history.

8.2D Θ $\mathfrak{F}(p)$ = [(#u)($p \equiv u$ says Θ p ω n a propn]
that fictively p

Something (somebody) says that p, which is not a proposition. But that p zu (see Volume 1), otherwise there would be nothing to say, and the only way this can happen is that that p is a predication. A historian can say that Poe or Dickens said that so-and-so. If that is all, then it is plain talk about an incident in the author's life. A presumpion of talking is that a person is affirming a proposition. Unless that is denied, as in 8.2D, one takes the assertion as true or false. Since it is cumbersome to repeat the denial that a proposition has been asserted, a discourse can be preceded by a blanket provision that an author's work and not his life is being discussed. Then, no matter who the author is, the predications discussed by the historian are fictional.